T0312660

Building Blocks for IoT Analytics

Internet-of-Things Analytics

RIVER PUBLISHERS SERIES IN SIGNAL, IMAGE AND SPEECH PROCESSING

The "River Publishers Series in Signal, Image and Speech Processing" is a series of comprehensive academic and professional books which focus on all aspects of the theory and practice of signal processing. Books published in the series include research monographs, edited volumes, handbooks and textbooks. The books provide professionals, researchers, educators, and advanced students in the field with an invaluable insight into the latest research and developments.

Topics covered in the series include, but are by no means restricted to the following:

- Signal Processing Systems
- Digital Signal Processing
- Image Processing
- Signal Theory
- Stochastic Processes
- Detection and Estimation
- Pattern Recognition
- Optical Signal Processing
- Multi-dimensional Signal Processing
- Communication Signal Processing
- Biomedical Signal Processing
- Acoustic and Vibration Signal Processing
- Data Processing
- Remote Sensing
- Signal Processing Technology
- Speech Processing
- Radar Signal Processing

For a list of other books in this series, visit www.riverpublishers.com

Building Blocks for IoT Analytics

Internet-of-Things Analytics

Editor

John Soldatos

Athens Information Technology
Greece

Published, sold and distributed by:
River Publishers
Alsbjergvej 10
9260 Gistrup
Denmark

River Publishers
Lange Geer 44
2611 PW Delft
The Netherlands

Tel.: +45369953197
www.riverpublishers.com

ISBN: 978-87-93519-03-9 (Hardback)
978-87-93519-04-6 (Ebook)

Contents

Preface xiii

List of Contributors xix

List of Figures xxi

List of Tables xxv

List of Abbreviations xxvii

PART I: IoT Analytics Enablers

1 Introducing IoT Analytics **3**
John Soldatos
1.1 Introduction . . . 3
1.2 IoT Data and BigData . . . 3
1.3 Challenges of IoT Analytics Applications . . . 5
1.4 IoT Analytics Lifecycle and Techniques . . . 7
1.5 Conclusions . . . 10
References . . . 10

2 IoT, Cloud and BigData Integration for IoT Analytics **11**
Abdur Rahim Biswas, Corentin Dupont and Congduc Pham
2.1 Introduction . . . 11
2.2 Cloud-based IoT Platform . . . 12
2.2.1 IaaS, PaaS and SaaS Paradigms . . . 12
2.2.2 Requirements of IoT BigData Analytics Platform . . . 13
2.2.3 Functional Architecture . . . 15
2.3 Data Analytics for the IoT . . . 15

2.3.1 Characteristics of IoT Generated Data 15
2.3.2 Data Analytic Techniques and Technologies 17
2.4 Data Collection Using Low-power, Long-range Radios . . . 20
2.4.1 Architecture and Deployment 20
2.4.2 Low-cost LoRa Implementation 21
2.5 WAZIUP Software Platform 23
2.5.1 Main Challenges 23
2.5.2 PaaS for IoT 24
2.5.3 Architecture 25
2.5.4 Deployment 26
2.6 iKaaS Software Platform 27
2.6.1 Service Orchestration and Resources Provisioning 30
2.6.2 Advanced Data Processing and Analytics 30
2.6.3 Service Composition and Decomposition 31
2.6.4 Migration and Portability in Multi-cloud Environment 33
2.6.5 Cost Function of Service Migration 35
2.6.6 Dynamic Selection of Devices in Multi-cloud Environment 35
Acknowledgement 36
References 37

3 Searching the Internet of Things **39**
Richard McCreadie, Dyaa Albakour, Jarana Manotumruksa, Craig Macdonald and Iadh Ounis
3.1 Introduction 39
3.2 A Search Architecture for Social and Physical Sensors 40
3.2.1 Search engine for MultimediA enviRonment generated contenT (SMART) 41
3.2.2 Challenges in Building an IoT Search Engine 46
3.3 Local Event Retrieval 48
3.3.1 Social Sensors for Local Event Retrieval 48
3.3.2 Problem Formulation 49
3.3.3 A Framework for Event Retrieval 51
3.3.4 Summary 53
3.4 Using Sensor Metadata Streams to Identify Topics of Local Events in the City 54

3.4.1 Definition of Event Topic Identification Problem 55
3.4.2 Sensor Data Collection 56
3.4.3 Event Pooling and Annotation 57
3.4.4 Learning Event Topics 59
3.4.5 Experiments 61
3.4.6 Summary 63
3.5 Venue Recommendation 63
3.5.1 Modelling User Preferences 65
3.5.2 Venue-dependent Evidence 67
3.5.3 Context-Aware Venue Recommendations 70
3.5.4 Summary 72
3.6 Conclusions 73
Acknowledgements 74
References 74

4 Development Tools for IoT Analytics Applications 81

John Soldatos and Katerina Roukounaki

4.1 Introduction 81
4.2 Related Work 82
4.3 The VITAL Architecture for IoT Analytics Applications . . . 84
4.4 VITAL Development Environment 87
4.4.1 Overview 87
4.4.2 VITAL Nodes 87
4.4.2.1 PPI nodes 88
4.4.2.2 System nodes 88
4.4.2.3 Services nodes 89
4.4.2.4 Sensors nodes 89
4.4.2.5 Observations nodes 89
4.4.2.6 DMS nodes 89
4.4.2.7 Query systems 89
4.4.2.8 Query services 89
4.4.2.9 Query sensors 89
4.4.2.10 Query observations 90
4.4.2.11 Discovery nodes 90
4.4.2.12 Discover systems nodes 90
4.4.2.13 Discover services nodes 90
4.4.2.14 Discover sensors nodes 90
4.4.2.15 Filtering nodes 90

4.4.2.16 Threshold nodes 90
4.4.2.17 Resample nodes 91
4.5 Development Examples 91
4.5.1 Example #1: Predict the Footfall! 91
4.5.2 Example #2: Find a Bike! 91
4.6 Conclusions 96
Acknowledgements 96
References 96

5 An Open Source Framework for IoT Analytics as a Service 99
John Soldatos, Nikos Kefalakis and Martin Serrano
5.1 Introduction 99
5.2 Architecture for IoT Analytics-as-a-Service 101
5.2.1 Properties of Sensing-as-a-Service Infrastructure 101
5.2.2 Service Delivery Architecture 102
5.2.3 Service Delivery Concept 105
5.3 Sensing-as-a-Service Infrastructure Anatomy 106
5.3.1 Lifecycle of a Sensing-as-a-Service Instance 106
5.3.2 Interactions between OpenIoT Modules 108
5.4 Scheduling, Metering and Service Delivery 112
5.4.1 Scheduler 112
5.4.2 Service Delivery & Utility Manager 118
5.5 Sensing-as-a-Service Example 122
5.5.1 Data Capturing and Flow Description 122
5.5.2 Semantic Annotation of Sensor Data 123
5.5.3 Registering Sensors to LSM 124
5.5.4 Pushing Data to LSM 125
5.5.5 Service Definition and Deployment Using OpenIoT Tools 126
5.5.6 Visualizing the Request 131
5.6 From Sensing-as-a-Service to IoT-Analytics- as-a-Service . . 134
5.7 Conclusions 136
Acknowledgements 137
References 137

6 A Review of Tools for IoT Semantics and Data Streaming Analytics 139
Martin Serrano and Amelie Gyrard
6.1 Introduction 139

6.2 Related Work 141
6.2.1 Linking Data 141
6.2.2 Real-time & Linked Stream Processing 142
6.2.3 Logic 142
6.2.4 Machine Learning 143
6.2.5 Semantic-based Distributed Reasoning 145
6.2.6 Cross-Domain Recommender Systems 146
6.2.7 Limitations of Existing Work 146
6.3 Semantic Analytics 147
6.3.1 Architecture towards the Linked Open Reasoning 148
6.3.2 The Workflow to Process IoT Data 149
6.3.3 Sensor-based Linked Open Rules (S-LOR) 152
6.4 Tools & Platforms 152
6.4.1 Semantic Modelling and Validation Tools 152
6.4.2 Data Reasoning 154
6.5 A Practical Use Case 156
6.6 Conclusions 157
Acknowledgement 157
References 158

PART II: IoT Analytics Applications and Case Studies

7 Data Analytics in Smart Buildings **167**
M. Victoria Moreno, Fernando Terroso-Sáenz, Aurora González-Vidal and Antonio F. Skarmeta
7.1 Introduction 167
7.2 Addressing Energy Efficiency in Smart Buildings 169
7.3 Related Work 174
7.4 A Proposal of General Architecture for Management Systems of Smart Buildings 179
7.4.1 Data Collection Layer 179
7.4.2 Data Processing Layer 180
7.4.3 Services Layer 181
7.5 IoT-based Information Management System for Energy Efficiency in Smart Buildings 181
7.5.1 Indoor Localization Problem 185
7.5.2 Building Energy Consumption Prediction 190

7.5.3 Optimization Problem 191
7.5.4 User Involvement in the System Operation 191
7.6 Evaluation and Results 192
7.6.1 Scenario of Experimentation 192
7.6.2 Evaluation and Indoor Localization Mechanism . . . 194
7.6.3 Evaluation. Energy Consumption Prediction and Optimization 195
7.6.4 Evaluation. User Involvement 197
7.7 Conclusions and Future Work 199
Acknowledgments 200
References 200

8 Internet-of-Things Analytics for Smart Cities 207

Martin Bauer, Bin Cheng, Flavio Cirillo, Salvatore Longo and Fang-Jing Wu

8.1 Introduction 207
8.2 Cloud-based IoT Analytics 208
8.2.1 State of the Art 209
8.3 Cloud-based City Platform 210
8.3.1 Use Case of Cloud-based Data Analytics 213
8.4 New Challenges towards Edge-based Solutions 215
8.5 Edge-based IoT Analytics 217
8.5.1 State of the Art 217
8.5.2 Edge-based City Platform 218
8.5.3 Workflow 221
8.5.4 Task and Topology 221
8.5.5 IoT-friendly Interfaces 222
8.6 Use Case of Edge-based Data Analytics 223
8.6.1 Overview of Crowd Mobility Analytics 223
8.6.2 Processing Tasks and Topology of Crowd Mobility Analytics 225
8.7 Conclusion and Future Work 226
References 227

9 IoT Analytics: From Data Collection to Deployment and Operationalization 231

John Soldatos and Ioannis T. Christou

9.1 Operationalizing Data Analytics Using the VITAL Platform 231

9.1.1 IoT Data Analysis 232
9.1.2 IoT Data Deployment and Reuse 232
9.2 Knowledge Extraction and IoT Analytics Operationalization 233
9.3 A Practical Example based on Footfall Data 234
Acknowledgement 237
References 237

10 Ethical IoT: A Sustainable Way Forward **239**
Maarten Botterman
10.1 Introduction 239
10.2 From IoT to a Data Driven Economy and Society 240
10.3 Way Forward with IoT 245
10.4 Conclusions 246
References 247

Epilogue **249**

Index **251**

About the Editor **253**

About the Authors **255**

Preface

The Internet-of-Things (IoT) is gradually being established as the new computing paradigm, which is bound to change the ways of our everyday working and living. IoT emphasizes the interconnection of virtually all types of physical objects (e.g., cell phones, wearables, smart meters, sensors, coffee machines and more) towards enabling them to exchange data and services among themselves, while also interacting with humans as well. Few years following the introduction of the IoT concept, significant hype was generated as a result of the proliferating number of IoT-enabled devices, which (according to many projections) are expected to amount to several billions in the next years. During recent years, this hype has been turning to reality, as a wave of IoT applications with significant social and economic has been emerging. This wave includes for example applications that deploy IoT devices and optimize operations associated with smart buildings, energy efficient neighborhoods, intelligent and sustainable urban transport systems, lifestyle management for disease prevention, management of water resources and more. In most cases, these applications rely on the processing of IoT data in order to optimize operations and facilitate decision making. Even in cases of IoT applications that deal with actuation and control (e.g., manufacturing robots), IoT data analysis plays a major role, as the foundation for driving the control process.

Overall, IoT data analysis is an integral element of any non-trivial IoT system. Nevertheless, IoT analytics are still in their infancy, as IoT data still remain largely unexploited. According to recent research reports and surveys, only 1% of IoT data are currently used, which is a serious set-back against IoT's business potential. The realization of IoT's expected (multi-trillion dollar) value requires the widespread deployment of IoT analytics applications, i.e. applications that collect and process data from multiple heterogeneous data sources. IoT analytics applications will ensure the proper exploitation of the proliferating volumes of IoT data for a variety of non-trivial business purposes, involving not only production of simple data-driven insights on operations, but also prediction of future trends and events. In this context, IoT stakeholders, including researchers, architects, application developers, solution integrators and service providers must be able to understand the challenges associated with the design, development, deployment and operation of advanced IoT analytics systems. Moreover, they are also expected to be able to identify and apply effective solutions to these challenges. The present book aims at contributing to this direction. In particular, it illustrates the challenges of developing, deploying and operationalizing IoT analytics applications, along with relevant solutions based on emerging future internet infrastructures and technologies, such as cloud computing and BigData. The authoring of the book is motivated by the need to provide IoT stakeholders with knowledge on how to confront the challenges of IoT analytics systems and applications, based on combinations of existing and emerging infrastructures and solutions, such as cloud and edge computing infrastructures, data streaming engines and databases, as well as BigData analytics infrastructures. In order to illustrate concrete solutions that successfully address the IoT analytics challenges, best practices stemming from successful application deployments are presented. On the whole, the book is structured in two complementary parts: one outlining technology enablers that empower non-trivial IoT analytics deployments and another one illustrating examples of IoT analytics deployments.

The "IoT analytics enablers" part of the book is structured around the main challenges of IoT analytics. These challenges include the great heterogeneity of the IoT data sources (including their diverse semantics), the typically high velocity of IoT data streams, the noisy and error-prone nature of IoT data, as well as the time and location dependent nature of IoT data resources. As a result of these characteristics, IoT analytics deployments require new solutions, notably solutions for handling streams with high ingestion rates, solutions for ensuring the semantic unification of IoT streams, novel techniques for reasoning over IoT data, as well as new tools for developing data-intensive applications. These solutions include elements from cloud computing and BigData technologies, given that IoT data streams feature several of the Vs of BigData (e.g., volume, variety, velocity, veracity) and can greatly benefit from the capacity, scalability, performance and pay-as-you-go nature of the cloud. However, in addition to the cloud and BigData elements, the presented IoT solutions include IoT technologies (e.g., for IoT data collection, preparation and semantic interoperability), which enable the adaptation of BigData analytics techniques to the IoT domain.

Overall, the first part of the book presents a wide range of popular tools that are used for the scalable collection and real-time processing of distributed IoT data streams such as: (a) The Apache Storm (storm.apache.org/), Spark (spark.apache.org/) and Flink (https://flink.apache.org/) open source projects and associated stream processing engines; (b) The OpenIoT (https://github.com/OpenIotOrg/openiot) open source toolkit for the semantic interoperability of heterogeneous IoT streams; (c) The Node-RED tool (nodered.org) for developing IoT applications based on the composition of IoT data flows; (d) A wide range of tools for semantic modeling and reasoning, which are discussed in a dedicated chapter of the "IoT analytics enablers" part of the book. Therefore, the first part of the book provides a comprehensive overview of some of the most popular tools for IoT data analytics, along with their use in practical projects and applications.

As far as IoT analytics applications and related best practices are concerned the book includes a set of representative applications. The latter are bundled in the second part of the book and include applications relating to smart buildings and crowd analysis in smart cities. Note that the smart city concept can serve as an umbrella for the presentation and discussion of different IoT analytics applications, as the latter can be deployed in the urban environment. It should be noted that the presented applications are by no means exhaustive or even representative of the broad scope of IoT analytics applications. Rather, they strive to provide concrete examples of application deployment, in order to facilitate the reader to understand practical issues in realistic environments. In addition to these applications, the ethical implications of IoT analytics in different application domains are discussed. The ethical implication chapter of the book, although not pure technological, complements the technology-related topics in terms of the prerequisites needed for IoT deployment.

It should be noted that this book represents a compilation of independent, yet interrelated chapters about different technologies and applications of IoT analytics. Several chapters of the book however cross-reference others, when pointers to technological topics (that are already discussed in the book) are required. All of the chapters' co-authors possess deep knowledge of IoT analytics issues, both at an academic and at a practical level, thus being excellent contributors to the present book. Most of the contributors have been speakers during the "IoT Analytics" session of the 5th IoT Week, which was held in Lisbon, back in June, 2015. The possibility of authoring an IoT analytics book was initially discussed following this session and led to the present book. Note also that several of the contributions are based on results of past and on-going research projects, which have been co-funded by the European Commission, as part of its FP7 and H2020 programmes. These projects are also acknowledged as important sources of the present book's contributions.

As already outlined the book is structured in two parts. The first one on "IoT Analytics Enablers" includes six contributions, in particular:

- An introductory chapter titled: "Introducing IoT Analytics", which is a natural extension of this preface and presents an overview of the main challenges of IoT analytics systems development and deployment.
- A second chapter titled "IoT, Cloud and BigData Integration for IoT Analytics", which is co-authored by Abdur Rahim Biswas, Corentin Dupont and Congduc Pham. This chapter illustrates the close affiliation between IoT, cloud computing and BigData technologies, including concrete system architectures for integrating them in real-life applications. Some of the presented results have been produced in the scope of EU projects WAZIUP and iKaaS.
- The third chapter of the book is titled "Searching the Internet-of-Things" and emphasizes on indexing and retrieval of data that stem from IoT devices. The chapter pays special emphasis in the presentation of techniques for high-performance in-memory indexing of IoT data, which is essential for data analysis (nearly) in real-time. It also illustrates the processing of social media streams as virtual IoT streams. Several of the discussed results have been produced in the scope of the FP7 SMART project.
- The presentation of development tools for IoT analytics is the main goal of the fourth chapter, which is titled "Development Tools for IoT Analytics Applications" and co-authored by Aikaterini Roukounaki and myself. It illustrates the limitations of state-of-the-art tools for IoT application development when it comes to implementing analytics applications and how they can be remedied based on existing frameworks for data analysis.
- The delivery of IoT analytics services is illustrated in the firth chapter of the book, which is co-authored by Nikos Kefalakis, Martin Serrano and myself. The title of the chapter is "An Open Source Framework for IoT Analytics-as-a-Service" and its content devoted to the presentation of a paradigm for on-demand IoT analytics based on the open source OpenIoT project.
- The importance of semantic modeling and semantic reasoning for IoT analytics is discussed in chapter six. This chapter is co-authored by Martin Serrano and Amelie Gyrard. It provides a comprehensive overview of tools and techniques for IoT semantic modeling and interoperability (including data linking), along with tools for semantic reasoning with particular emphasis on the tools that are applicable to IoT data streams.

At the same time the second part of the book, emphasizes on applications examples and case studies as follows:

- The seventh chapter of the book (and first of the book's second part) is devoted to presenting a system for analyzing data in smart buildings, including data stemming from sensors and IoT devices. The chapter is titled "Data Analytics in Smart Buildings" and co-authored by M. Victoria Moreno, Fernando Terroso-Sáenz, Aurora González-Vidal, and Antonio F. Skarmeta.
- The eighth chapter of the book is titled "IoT analytics for Smart Cities" and presents a case study concerning IoT analytics deployments in urban environments, with particular emphasis on cloud analytics applications. It is co-authored by Martin Bauer, Bin Cheng, Flavio Cirillo, Salvatore Longo and Fang-Jing Wu.
- The process of deploying advanced machine learning techniques in the scope of an IoT analytics application is presented in the ninth chapter of the book, which is co-authored by Ioannis Christou and myself. The chapter is titled "IoT Analytics: From Data Collection to Deployment and Operationalization" and illustrates the lifecycle of an IoT analytics applications across all the required stages, including data collection, data interoperability and deployment of appropriate data mining and/or machine learning schemes for the analytics problem at hand. The chapter is linked to Chapter 4, given that both refer to the IoT-based smart city platform which has been developed in the scope of the VITAL project.
- The tenth and last chapter of the book is co-authored by Maarten Botterman and titled "Ethical IoT: a sustainable way forward". It underlines the ethical challenges that are associated with IoT analytics applications, along with best practices for successfully confronting them.

Overall this is one of the first and few books to discuss advanced IoT analytics topics, including relevant technologies and case studies. I therefore believe it could provide insights on IoT analytics to interested parties for the coming years, while I also expect that additional books on IoT analytics will also emerge. I sincerely hope that readers will find the book interesting and worth reading.

John Soldatos
July, 2016

List of Contributors

Abdur Rahim Biswas, *CREATE-NET, Italy*

Amelie Gyrard, *Insight Center for Data Analytics, National University of Ireland, Galway, Ireland*

Antonio F. Skarmeta, *Department of Information and Communications Engineering, University of Murcia, 30100 Spain*

Aurora González-Vidal, *Department of Information and Communications Engineering, University of Murcia, 30100 Spain*

Bin Cheng, *NEC Laboratories Europe, Heidelberg, Germany*

Congduc Pham, *University of Pau, France*

Corentin Dupont, *CREATE-NET, Italy*

Craig Macdonald, *University of Glasgow, UK*

Dyaa Albakour, *Signal Media, UK*

Fang-JingWu, *NEC Laboratories Europe, Heidelberg, Germany*

Fernando Terroso-Sáenz, *Department of Information and Communications Engineering, University of Murcia, 30100 Spain*

Flavio Cirillo, *NEC Laboratories Europe, Heidelberg, Germany*

Iadh Ounis, *University of Glasgow, UK*

Ioannis T. Christou, *Athens Information Technology, Greece*

Jarana Manotumruksa, *University of Glasgow, UK*

John Soldatos, *Athens Information Technology, Greece*

Katerina Roukounaki, *Athens Information Technology, Greece*

M. Victoria Moreno, *Department of Information and Communications Engineering, University of Murcia, 30100 Spain*

Maarten Botterman, *GNKS Consult NV, Netherlands*

Martin Bauer, *NEC Laboratories Europe, Heidelberg, Germany*

Martin Serrano, *Insight Center for Data Analytics, National University of Ireland, Galway, Ireland*

Nikos Kefalakis, *Athens Information Technology, Greece*

Richard McCreadie, *University of Glasgow, UK*

Salvatore Longo, *NEC Laboratories Europe, Heidelberg, Germany*

List of Figures

Figure 1.1	The Vs of BigData and IoT (Big)Data.	4
Figure 1.2	Main challenges and lifecycle phases of IoT analytics.	7
Figure 2.1	Functional Architecture of IoT and Bigdata platform.	16
Figure 2.2	BigData properties.	17
Figure 2.3	IoT BigData applications.	18
Figure 2.4	Gateway-centric deployment.	21
Figure 2.5	Low cost gateway from off-the-sheves components.	22
Figure 2.6	PaaS deployment extended for IoT in WAZIUP.	25
Figure 2.7	WAZIUP architecture.	26
Figure 2.8	WAZIUP local and global deployment.	27
Figure 2.9	iKaaS platform.	29
Figure 2.10	Service composition and decomposition.	32
Figure 2.11	Patten for composition and decomposition.	33
Figure 2.12	iKaaS distributed local and global cloud with service migration.	34
Figure 3.1	Data sources available to an IoT-connected search engine.	40
Figure 3.2	Architecture of the SMART framework.	42
Figure 3.3	Edge node components.	43
Figure 3.4	Components overview of the SMART search layer.	45
Figure 3.5	A plot of the volume of tweets in London that contain the phrase "beach boys" over time.	48
Figure 3.6	Components a snapshot from the annotation interface.	58
Figure 3.7	Word clouds for frequent terms occurring in the English (a) and Spanish (b) annotations, respectively, where word size is indicative of frequency in the annotations.	59

Figure 3.8 Obtaining DMOZ category distributions from Facebook' likes and venues. 67
Figure 3.9 Predicting occupancy from Foursquare check-in time-series. 68
Figure 3.10 Percentage of improvement obtained when independently removing single venue-dependent features, with respect to a LambdaMART baseline that uses a total of 64 features. Improvements are expressed in terms of P@5, P@10, and MRR. Statistical significance is stated according to a paired t-test (*: $p < 0.05$, **: $p < 0.01$, ***: $p < 0.001$). 69
Figure 3.11 Screenshots of the EntertainMe! mobile venue recommendation application. 73
Figure 4.1 VITAL platform architecture. 85
Figure 4.2 Elements of the VITAL development tool. 88
Figure 4.3 Predict the footfall – the web page. 92
Figure 4.4 Predict the footfall – the flows. 93
Figure 4.5 Find a bike – the web page. 94
Figure 4.6 Find a bike – the flows. 95
Figure 5.1 OpenIoT architecture. 103
Figure 5.2 Functional blocks of openIoT's project analytics as a service architecture. 104
Figure 5.3 IoT data analysis services request lifecycle. 107
Figure 5.4 Main entities and modules. 109
Figure 5.5 Relationships between the main OpenIoT data entities. 112
Figure 5.6 State diagram of the OpenIoT services lifecycle within the scheduler module. 114
Figure 5.7 "Register Service" process flowchart. 117
Figure 5.8 "Update Resources" service flowchart. 119
Figure 5.9 "Unregister" service flowchart. 120
Figure 5.10 Request definition log in. 126
Figure 5.11 Request definition loaded profile. 127
Figure 5.12 New application creation. 127
Figure 5.13 Sensor discovery in Brussels area. 128
Figure 5.14 Comparator (between) properties. 129
Figure 5.15 Grouping options. 130
Figure 5.16 Line chart properties. 130

Figure 5.17 Validation of the service design. 131
Figure 5.18 SPARQL script generation. 132
Figure 5.19 LSM SPARQL endpoint (2 weeks wind chill in Brussels). 132
Figure 5.20 Save application button. 133
Figure 5.21 Request presentation loaded profile. 133
Figure 5.22 Load "WeatherInBrussels" scenario. 134
Figure 5.23 Wind chill vs. air temperature in Brussels line chart. 135
Figure 6.1 Summary of existing approaches for IoT data enrichment. 147
Figure 6.2 Classification of tools according to reasoning approaches. 148
Figure 6.3 Reasoning main operations. 149
Figure 6.4 IoT process defined by SEG 3.0 methodology. . . . 150
Figure 6.5 IoT reasoning data framework within FIESTA-IoT. 156
Figure 7.1 Layers of the base architecture for smart buildings ecosystem. 180
Figure 7.2 City explorer applied to smart buildings. 182
Figure 7.3 Schema of the modules composing the management system in charge of the building comfort and energy efficiency. 184
Figure 7.4 Outline of some positioning technologies. 186
Figure 7.5 Localization scenario. 187
Figure 7.6 Data processing for location estimation. 188
Figure 7.7 Schema of the definitive module of our building energy management system. 193
Figure 7.8 Tracking processes with a reference tag distribution of 1 m × 1 m. 194
Figure 7.9 Modeling results. 196
Figure 7.10 Percentage of energy consumption savings in comfort services considering a user-centric building management efficient. 198
Figure 8.1 System architecture of the CiDAP platform. 211
Figure 8.2 Subscription mechanisms to get real-time notifications. 213
Figure 8.3 System overview of the cloud-based analytics. . . . 214

Figure 8.4 Subscription mechanisms to get real-time notifications. 219
Figure 8.5 Task topology and processing topology. 222
Figure 8.6 A system overview of the edge-based data analytics for crowd mobility. 224
Figure 9.1 Camden footfall dataset. 235
Figure 10.1 From IoT to BigData and analytics. 240

List of Tables

Table 3.1 Summary of sensor data collection 57
Table 3.2 Statistics from annotated videos 58
Table 3.3 Topics identifed with topic modelling using the english annotations 60
Table 3.4 Distribution of labels 60
Table 3.5 Features devised for topics identification 61
Table 3.6 Performance of topic identification 62
Table 3.7 Results of the ablation study 62
Table 3.8 Sources of data for venue recommendation 64
Table 3.9 Examples of venue recommendations produced by our model for user in a central location in London at two different times 68
Table 3.10 Venue-dependent Foursquare features used by Deveaud et al. 70
Table 3.11 The 12 dimensions of the contextual aspects proposed by the TREC 2015 contextual suggestion track . . . 71

List of Abbreviations

AI	Artificial Intelligence
ASHRAE	American Society of Heating, Refrigerating, and Air-Conditioning Engineers
AIOTI	Alliance for Internet Of Things Innovation
API	Application Programming Interface
BMS	Building Management System
BRNN	Bayesian Regularized Neural Networks
BSSID	Basic Service Set Identifier
CAPEX	Capital Expenses
CAVR	Context-aware venue recommendation
CCTV	Close Circuit TeleVision
CEN	European Committee for Standardization
CEP	Complex Event Processing
CF	Collaborative filtering
CiDAP	City Data and Analytics Platform
CO2	Carbon dioxide
CRISP-DM	Cross Industry Standard Process for Data Mining
CVRMSE	Coefficient of variation of the Root-Mean-Square Error
DaaS	Data as a Service
DB	Database
DIFS	Distributed Interframe Space
DMOZ	Open Directory Project
DMS	Data Management Service
EC	European Commission
EDPS	European Data Protection Supervisor
ETL	Extract Transofm Load
EU	European Union
FCAPS	Fault Configuration Accounting Performance and Security
FP7	7th Framework Programme of the European Commission
FVEY	FiVE Eyes
GA	Genetic Algorithm

GDPR	General Data Protection Regulation
Geelytics	Geo-distributed edge analytics
GHG	Green House Gas
GIS	Geographical Information System
GSM	Global System for Mobile communications
GSN	Global Sensor Networks
GUI	Graphical User Interface
HAM	Home Automation Module
HDFS	Hadoop Distributed File System
HTTP	Hypertext Transfer Protocol
HVAC	Heating, Ventilating, and Air Conditioning
H2020	Horizon 2020
IaaS	Infrastructure as a Service
IBMS	Intelligent Building Management System
ICE	Integrated Cloud Environment
ICT	Information and Communications Technology
ICO	Internet Connected Objects
IDE	Integrated Development Environment
Idir	IR identifier value
IGF DC IoT	Internet Governance Forum's Dynamic Coalition of the Internet of Things
I/O	Input/Output
IoT	Internet of Things
IR	Infra-Red
IR	Information Retrieval
JFK	John F. Kennedy Airport New York
JSON	JavaScript Object Notation
JSON-LD	JavaScript Object Notation Linked Data
KAT	Knowledge Acquisition Toolkit
KL	Kullback Leibler
LAN	Local Area Network
LBSN	Location-based social network
LDA	Latent Dirichlet Allocation
LER	Linked Edit Rules
LOD	Linked Open Data
LOR	Linked Open Rules
LoRa	LoRa
LOS	Line of Sight
LOV	Linked Open Vocabularies

LOV4IoT	Linked Open Vocabularies for Internet of Things
LPWAN	Low-Power Wide Area Networks
MAC	Media Access Control
MALLET	MAchine Learning for LanguagE Toolkit
MART	Multiple Additive Regression Trees
MDP	Markov Decision Process
MF	Matrix Factorisation
MLP	Multilayer Perceptron
MQTT	Message Queuing Telemetry Transport
MRR	Mean Reciprocal Rank
M2M	Machine to Machine
NGSI	Next Generation Service Interfaces
NILM	Non-Intrusive Load Monitoring
NIST	National Institute of Standards & Technology
NLOS	Non Line of Sight
NSA	National Security Agency
OGC	Open Geospatial Consortium
OMA	Open Mobile Alliance
OPEX	Operational Expenses
PaaS	Platform as a Service
PADA	Platform Agnostic Data Access
PCA	Principal Component Analysis
PF	Particle Filter
POI	Point of Interest
PPI	Platform Provider Interface
Pub/Sub	Publish Subscribe
QD	Query Dependent
QI	Query Independent
QoS	Quality of Service
QARM	Quantitative Association Rule Mining
QARMA	Quantitative Association Rule Mining Algorithm
RAD	Rapid Application Development
RBF	Radial Basis Function
RDBMS	Relational Database Management Systems
REST	Representational State Transfer
RF	Radio Frequency
RFID	Radio Frequency Identification
RMSE	Root-Mean-Square Error
SaaS	Software as a Service

SAN	Sensor Actuator Network
SD	Service Discovery
SHA	Secure Hashing Algorithm
SIFS	Short Interframe Space
SIS	Sequential importance sampling
SLA	Service Level Agreement
SMART	Search engine for MultimediA enviRonment generated contenT
SME	Small to Medium Enterprise
SPARQL	SPARQL Query Language
SQL	Structured Query Language
SSN	Semantic Sensor Networks
SVN	Support Vector Machine
SWRL	Semantic Web Rule Language
S-LOR	Sensor-based Linked Open Rules
TDT	Topic Detection and Tracking
TREC	Text REtrieval Conference
TRECVID	A spinout of TREC focussing upon content-based retrieval and exploitation of digital VIDeo
TTC	Technology Transfer Centre
UAV	Unmanned Aerial Vehicles
UMU	University of Murcia
URL	Uniform Resource Locator
USA	United States of America
VM	Virtual Machine
VUAI	Virtualized Unified Access Interface
WoT	Web of Things
WSN	Wireless Sensor Network

PART I

IoT Analytics Enablers

1

Introducing IoT Analytics

John Soldatos

Athens Information Technology, Greece

1.1 Introduction

The internet-of-things (IoT) paradigm represents one of the next evolutionary steps in internet-based computing, which is already having a positive impact in a large number of application domains including smart cities, sustainable living, healthcare, manufacturing and more. IoT analytics refers to the analysis of data from multiple IoT data sources, including sensors, actuators, smart devices and other internet connected objects. The collection and analysis of data streams from IoT sources is nowadays considered a key element of the IoT's disruptive power, as well as a prerequisite to realizing IoT's hyped market potential. Indeed, according to a recent report by McKinsey [1], less than 1% of IoT data is currently used, which is a serious set-back to maximizing IoT's business value. For example, most IoT analytics applications are nowadays used for anomaly detection and control rather than for optimization and prediction, which are the applications that will provide the greatest business value in the coming years.

1.2 IoT Data and BigData

The rise of future internet technologies, including cloud computing and BigData analytics, enables the wider deployment and use of sophisticated IoT analytics applications, beyond simple sensor processing applications. It is therefore no accident that IoT technologies are converging with cloud computing and BigData analytics technologies towards creating and deploying advanced applications that process IoT streams.

The integration of IoT data streams within cloud computing infrastructures enables IoT analytics applications to benefit from the capacity, performance

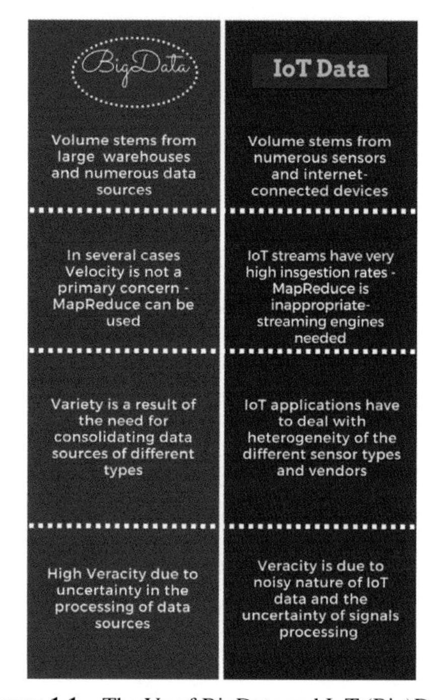

CONVENTIONAL
BIGDATA
VERSUS
IOT (BIG)DATA
COMPARING & UNDERSTANDING THEIR VS
BigData
IoT Data
Volume stems from large warehouses and numerous data sources
Volume stems from numerous sensors and internet-connected devices
In several cases Velocity is not a primary concern - MapReduce can be used
IoT streams have very high insgestion rates - MapReduce is inappropriate- streaming engines needed
Variety is a result of the need for consolidating data sources of different types
IoT applications have to deal with heterogeneity of the different sensor types and vendors
High Veracity due to uncertainty in the processing of data sources
Veracity is due to noisy nature of IoT data and the uncertainty of signals processing

Figure 1.1 The Vs of BigData and IoT (Big)Data.

and scalability of cloud computing infrastructures. In several cases, IoT analytics applications are also integrated with edge computing infrastructures, which decentralize processing of IoT data streams at the very edge of the network, while transferring only selected IoT data from the edge devices to the cloud. Therefore, it is very common to deploy IoT analytics applications within edge and/or cloud computing infrastructures.

In addition to the affiliation between IoT analytics and cloud computing infrastructures, there is a close relation between IoT analytics with BigData analytics. Indeed, IoT data are essentially BigData since they feature several of the Vs of BigData, including (Figure 1.1):

- **Volume**: IoT data sources (such as sensors) produce in most cases very large volumes of data, which typically exceed the storage and processing capabilities of conventional database systems.
- **Velocity**: IoT data streams have commonly very high ingestion rates, as they are produced continually, in very high frequencies and in several times in very short timescales.
- **Variety**: Due to the large diversity of IoT devices, IoT data sources can be very heterogeneous both in terms of semantics and data formats.
- **Veracity**: IoT data are a classical example of noise data, which are characterized by uncertainty.

Therefore, systems, tools and techniques for developing and deploying BigData applications (including databases, data warehouses, streaming middleware and engines, data mining techniques and BigData developments tools), provide a good starting point for dealing with IoT analytics. However, IoT data and IoT analytics applications have in most cases to deal with their own peculiar challenges, which are not always common to the challenges of high volume, high velocity transactional applications. The tools and techniques that are discussed in this book are focused on the challenges of IoT data and IoT analytics applications, which are outlined in the following paragraph.

1.3 Challenges of IoT Analytics Applications

The main challenges associated with the development and deployment of IoT analytics applications are (Figure 1.2):

- **The heterogeneity of IoT data streams**: IoT data streams tend to be multi-modal and heterogeneous in terms of their formats, semantics and velocities. Hence, IoT analytics applications expose typically variety and veracity. BigData technologies provide the means for dealing with this heterogeneity in the scope of operationalized applications. However, accessing IoT data sources (including sensors and other types of internet connected devices) requires drivers and connectors, beyond what is typically deployed in transactional BigData applications (e.g., database drivers). Furthermore, dealing with semantic interoperability of diverse data streams requires techniques beyond the (syntactic) homogenization of data formats.
- **The varying data quality**: Several IoT streams are noisy and incomplete, which creates uncertainty in the scope of IoT analytics applications.

Statistical and probabilistic approaches must be therefore employed in order to take into account the noisy nature of IoT data streams, especially in cases where they stem from unreliable sensors. Also, different IoT data streams can be typically associated with different reliability, which should be considered in the scope of their integration in IoT analytics applications.

- **The real-time nature of IoT datasets**: IoT streams feature high velocities and for several application must be processed nearly in real-time. Hence, IoT analytics can greatly benefit from data streaming platforms, which are part of the BigData ecosystem. IoT devices (e.g., sensors) provide typically high-velocity data, which however can be in several cases controlled by focusing only on changes in data patterns and reports, rather than dealing with all the observations that stem from a given sensor.
- **The time and location dependencies of IoT streams**: IoT data come with temporal and spatial information, which is directly associated with their business value in a given application context. Hence, IoT analytics applications must in several cases process data in a timely fashion and from proper locations. Cloud computing techniques (including edge computing architectures) can greatly facilitate timely processing of information from given locations in the scope of large scale deployments. Note also that the spatial and temporal dimensions of IoT data can serve as a basis for dynamically selecting and filtering streams towards analytics applications for certain timelines and locations.
- **Privacy and security sensitivity**: IoT data are typically associated with stringent security requirements and privacy sensitivities, especially in the case of IoT applications that involve the collection and processing of personal data. Hence, IoT analytics need to be supported by privacy preservation techniques, such as the anonymization of personal data, as well as techniques for encrypted and secure data storage.
- **Data bias**: As in the majority of data mining problems, IoT datasets can lead to biased processing and hence a thorough understanding and scrutiny of both training and test datasets is required prior to their operationalized deployment. To this end, classical data mining techniques can also be applied in the IoT case. Note that the specification and deployment of IoT analytics systems entails techniques similar to those deployed in classical data mining problems, including the understanding of the data, the preparation of the data, the testing of data mining techniques and ultimately the development and deployment of a system that yields the desired performance and efficiency.

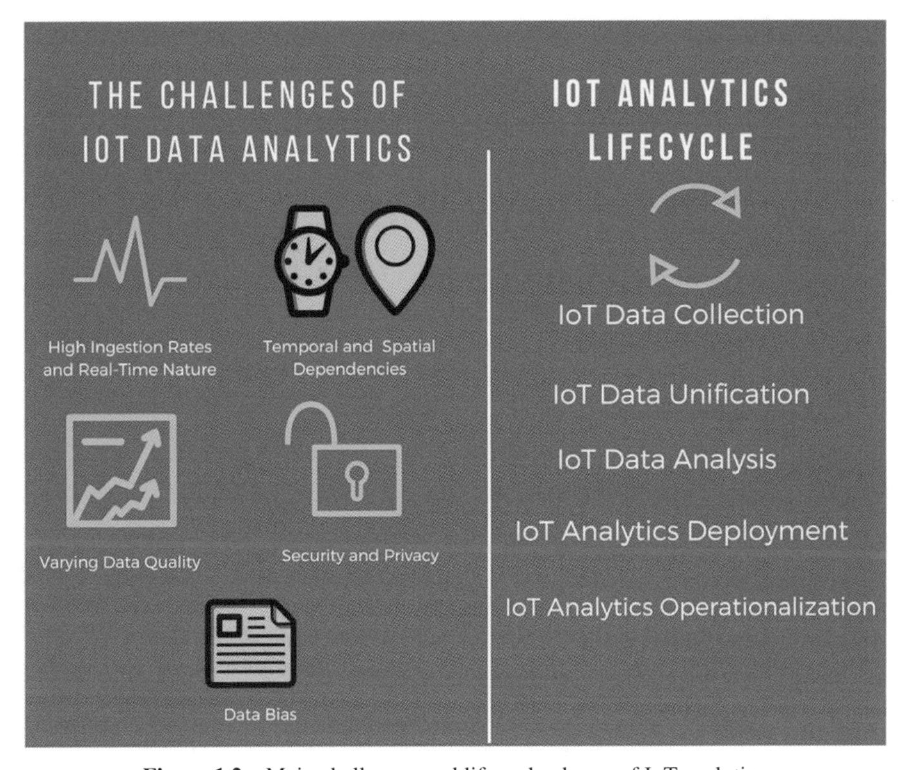

Figure 1.2 Main challenges and lifecycle phases of IoT analytics.

These challenges are evident in the IoT analytics lifecycle, which comprises a series of steps from data acquisition to analysis and visualization. This lifecycle is supported by cloud computing and BigData technologies, including data mining, statistical computing and scalable databases technology.

1.4 IoT Analytics Lifecycle and Techniques

The IoT analytics lifecycle comprises the phases of data collection, analysis and reuse. In particular:

- **1st Phase – IoT Data Collection**: As part of this phase IoT data are collected and enriched with the proper contextual metadata, such as location information and timestamps. Moreover, the data are validated in terms of their format and source of origin. Also, they are validated in terms of their integrity, accuracy and consistency. Hence, this phase addresses several IoT analytics challenges, such as the need to ensure consistency and quality. Note that IoT data collection presents several peculiarities, when compared to traditional data consolidation of distributed data sources, such as the need to deal with heterogeneous IoT streams.

- **2nd Phase – IoT Data Analysis**: This phase deals with the structuring, storage and ultimate analysis of IoT data streams. The latter analysis involves the employment of data mining and machine learning techniques such as classification, clustering and rules mining. These techniques are typically used to transform IoT data to actionable knowledge.
- **3rd Phase – IoT Data Deployment, Operationalization and Reuse**: As part of this phase, the IoT analytics techniques identified in the previous steps are actually deployed, thus becoming operational. This phase ensures also the visualization of the IoT data/knowledge according to the needs of the application. Moreover, it enables the reuse of IoT knowledge and datasets across different applications.

These lifecycle phases are used in order to organize the development and deployment of IoT analytics systems. They can also serve as a basis for characterizing the maturity of IoT analytics deployments. As a prominent example, they can be used to analyze the level of "smartness" of a city, given that the maturity of a city is directly related to the sophistication of its analytics, but also to its ability to repurpose and reuse datasets and data analytics functions.

The tasks outlined in the above listed phases are supported by a range of data management and analysis disciplines, including:

- **IoT middleware and interoperability technologies**, which provide the means for collecting, structuring and unifying IoT data streams, thus addressing the variety and veracity challenges of IoT data.
- **Statistics**, which provide the theory for testing hypotheses about various insights stemming from IoT data.
- **Machine learning**, which enables the implementation of learning agents based on IoT data mining. Machine learning includes several heuristic techniques. The practical cases studies at the second part of the book make use of various machine learning schemes.
- **Data mining and Knowledge Discovery**, which combines theory and heuristics towards extracting knowledge. To this end, data cleaning, learning and visualization might be also employed.
- **Database management systems**, including Relational Database Management Systems (RDMS), NoSQL databases, BigData databases (such

as the HDFS (Hadoop Distributed File System), which provide the means for data persistence and management. Most of the practical examples and case studies presented in the book make use of some sort of database management systems in order to persist and manage the data.

- **Data streams management systems**, which handle transient streams, including continuous queries, while being able to handle data with very high ingestion rates, including streams featuring unpredictable arrival times and characteristics. IoT streaming systems are also supported by scalable, distributed data management systems.

The techniques that are outlined as part of subsequent chapters of this book use and in several cases enhance the above-listed techniques and systems for data collection, management and analysis. For example, the following chapters make direct references to distributed real-time streaming and event processing systems like Apache Spark[1] and Apache Storm[2]. Apache Storm is a free and open source distributed real-time computation system. It facilitates reliable processing of unbounded streams of data and deals with Real-time processing much in the same way Apache Hadoop deals with batch processing. Similarly, Apache Storm is an open source software that defines a broader set of operations when comparing to Hadoop, including transformation and actions which can be arbitrarily combined in any order. Spark supports several programming Languages including Java, Scala and Python. Note that the choice between Spark or Storm for IoT streaming and analytics can be based on a number of different factors. Spark is usually a good choice for projects using existing Hadoop or Mesos clusters, as well as for projects involving considerable graph processing, SQL access, or batch processing. Moreover, Spark provides a shell for interactive processing (something missing from Storm). On the other hand, Storm is a good choice for projects primarily focused on stream processing and Complex Event Processing that have structures matching Storm's capabilities. Storm provides boader language support, including support for the R language which is extremely popular among data scientists. Beyond Apache Spark and Storm projects, Apache Flink[3] is another open source stream processing framework, which can support low latency, high throughput, stateful and distributed processing for IoT data. It provides low-latency streaming ETL (Extract-Transform-Load) operations, offering much higher performance than traditional ETL for batch datasets. Moreover,

[1] spark.apache.org

[2] storm.apache.org

[3] flink.apache.org

Flink is event-time aware: Events stemming from the same real-world activity could arrive out of order in a Flink streaming system, but even in such cases Flink can maintain the order. Flink is a more recent project comparing to Spark, but it constantly gaining momentum in the industry due to its innovative and high-performance functionalities. Likewise the applications that are illustrated in the second part of the book take also advantage of these techniques. For example, several of the presented applications exploit NoSQL databases (such as MongoDB[4] and CouchDB[5] for data storage and management, while most of the applications deploy also some data mining method like classification, prediction and mining of association rules.

1.5 Conclusions

This introductory chapter has defined the scope of IoT analytics and presented related technologies. It has also outlined the close affiliation of IoT analytics with the cloud computing and BigData techniques. Furthermore, it has presented the main challenges of IoT analytics applications, which stem primarily from the unique characteristics and nature of IoT data. The rest of the book is destined to present technology solutions to these challenges, along with practical applications and case studies, which make use of such solutions. The presented solutions build in several cases over state-of-the-art IoT, cloud computing and BigData solutions, given that the integration of these technologies tends to become a norm for the variety of IoT analytics applications. The integration of IoT, cloud computing and BigData infrastructures and technologies is therefore the topic discussed in the next chapter.

[4]https://www.mongodb.com/

[5]http://couchdb.apache.org/

References

[1] Manyika, J., Chui, M., Bisson, P., Woetzel, J., Dobbs, R. Bughin, J., Aharon, D. *Unlocking the Potential of the Internet of Things*. McKinsey Global Institute, June (2015).

[2] J. Soldatos, et al. "*IoT analytics: Collect, Process, Analyze, and Present Massive Amounts of Operational data – Research and Innovation Challenges*" Chapter 7 in Book, "Building the Hyperconnected Society – IoT Research and Innovation Value Chains, Ecosystems and Markets", IERC Cluster Book 2015, River Publishers.

2

IoT, Cloud and BigData Integration for IoT Analytics

Abdur Rahim Biswas[1], Corentin Dupont[1] and Congduc Pham[2]

[1]CREATE-NET, Italy
[2]University of Pau, France

2.1 Introduction

Over the last years, the Internet of Things (IoT) has moved from being a futuristic vision to market reality. It is not a question any more whether IoT will be surpassing the hype, it is already there and the race between IoT industry stakeholders has already begun. The IoT revolution comes with trillions of connected devices; however the real value of IoT is in the advanced processing of the collected data. By nature, IoT data is more dynamic, heterogeneous and unstructured than typical business data. It demands more sophisticated, IoT-specific analytics to make it meaningful. The exploitation in the Cloud of data obtained in real time from sensors is therefore very much a necessity. This data processing leads to advanced proactive and intelligent applications and services. The connection of IoT and BigData can offer: i) deep understanding of the context and situation; ii) real-time actionable insight; iii) performance optimization; and iv) proactive and predictive knowledge. Cloud technologies offer decentralized and scalable information processing and analytics, and data management capabilities. This chapter describes a Cloud based IoT and BigData platform, together with their requirements. This includes multiple sensors and devices, BigData analytics, cloud data management, edge-heavy computing, machine learning and virtualization.

In this chapter, Section 2.2 introduces the characteristics of an online Cloud IoT platform. Section 2.3 shows the challenge posed by the huge amount of data to be processed, from the point of view of the quality and quantity of data. It gives an overview of the technologies able to address those challenges.

Section 2.4 presents LoRa, a key enabler for the collection of the data. The chapter includes also initial results of two EU-funded projects on IoT BigData: WAZIUP in Section 2.5; and iKaaS in Section 2.6.

2.2 Cloud-based IoT Platform

According to the NIST definition, Cloud computing is a model for enabling convenient, on-demand network access to a shared pool of configurable computing resources that can be rapidly provisioned and released with minimal management effort or service provider interaction. The Cloud paradigm can be delivered using essentially three different service models. These are Infrastructure as a Service (IaaS), Platform as a Service (PaaS), and Software as a Service (SaaS).

A Cloud-based IoT platform is then a dynamic and flexible resource sharing platform delivering IoT services. It offers scalable resources and services management. The exploitation of IoT data depends on massive resources, which should be available when needed and scaled back when not needed.

2.2.1 IaaS, PaaS and SaaS Paradigms

A Cloud based IoT platform needs usually to select one from the three different service models: IaaS, PaaS or SaaS. IaaS allows delivering computer infrastructure on an outsourced basis in order to support enterprise operations. This service model is based on the paradigm of virtualization of resources. The initial success of the Cloud is due to the possibility to embed practically any legacy applications within Virtual Machines (VMs), which are managed by an external stakeholder. This permits to relieve the application owner from managing physical infrastructures. PaaS, on the other hand, provides a platform allowing customers to develop, run, and manage applications. It removes the complexity of building and maintaining the infrastructure typically associated with developing and deploying an application. Typically, a PaaS framework will compile an application from its source code, and then deploy it inside lightweight VMs, or containers. Furthermore, PaaS environments offer an interface to scale up or down applications, or to schedule various tasks within the applications. Finally, SaaS is a software licensing and delivery model in which software is licensed on a subscription basis and is centrally hosted. It is sometimes referred to as "on-demand software". SaaS is typically accessed by users using a thin client via a web browser.

Cloud-based IoT platforms are usually based on the SaaS paradigm. They provide IoT-related services using a web interface on a pay-per-use basis. For example, a service such as Xively[1] provides a web service with a database able to store sensors data points. This data is then processed and displayed in various graphics.

However, SaaS IoT platforms are limited to the possibility of their web interface. They will not permit the developers to create complex and custom applications. Extensibility mechanisms are sometime offered, allowing extending the web services offered with user-provided callbacks. However the resulting application will not be homogeneous and will be difficult to maintain. Instead, we present in Section 2.5 a concept of IoT Cloud platform based on the PaaS paradigm. Developing an IoT BigData application is a complex task. A lot of services need to be installed and configured, such as databases, message broker and big data processing engines. With the PaaS paradigm, we abstract some of this work. The idea is to let the developer specify the requirements of his application in a specification file called the "manifest". This specification will be read by the PaaS framework and the application will be compiled and instantiated in the Cloud environment, together with its required services.

2.2.2 Requirements of IoT BigData Analytics Platform

An IoT BigData analytic platform should be able to dynamically manage IoT data and provide connectivity with the diverse heterogeneous objects, considering the interoperability issues. It is able to derive useful information and knowledge from large volume of IoT data. The platform shall offer ubiquitous accessibility and connectivity of the diverse objects, services and users, in a mobile context. It shall allow dynamic management and orchestration of users, a huge amount of connected devices as well as massive amount of data produced by those devices. Finally it shall allow personalization of users and services, providing services based on users preference and requirements including real-world context.

Intelligent and Dynamic

The platform should include intelligent and autonomic features in order to dynamically manage the platform functions, components and applications. The platform should also be capable to make proactive decisions, dynamic deployment, and intelligent decisions based on the understanding of the

[1] https://xively.com

context of the environments, users and applications requirements. The platform provides dynamic resources management for IoT, considering performance targets and constraints. This includes offloading workload from clients/hosts to the Cloud and dynamic resources and service migration, as presented in Section 2.6.

Distributed

The platform includes distributed information processing and computing capabilities, distributed storage, distributed intelligence, and distributed data management capabilities. These capabilities should be distributed across smart devices, gateway/server and multiple cloud environments. The processing capability needs to be migrated closer to users, to save bandwidth.

Scalable

The platform needs to be scalable in order to address the needs of a variable number of the devices, services and users. The data management, storage and processing services need to be dimensioned dynamically.

Real-Time

The platform need to be able to process data in real-time, i.e. providing a fast analysis and responses for situations of urgency. A real-time data analysis platform needs to be able to prioritize urgent traffic and processing from non-urgent ones.

Programmable

The platform shall support programmable capabilities of IoT business and service logics, data warehouse scheme, template of data and service model.

Interoperable

The platform provides interoperability between the different IoT services and infrastructure. The APIs need to follow the existing standards. The components are published and maintained as Open Source software. The target is to deliver a common data model able to exploit both structured and unstructured data. In order to create multimodal and cross-domain smart applications, it is necessary to move from raw data to linked data and adopt unambiguous description of relevant information.

Secure

The platform shall include security and privacy by design. This includes different features like data integrity, localization, confidentiality, SLA. Holistic approaches are required to address privacy & security issues across value chains.

2.2.3 Functional Architecture

Our IoT platform solves key problems in IoT analytics, data management and visualization that have traditionally been developed within each application. Developers can easily embed the platform components into their applications saving the time, expertise and expense of building the components themselves. This enables application that would have been too costly and time-consuming to develop. The platform integrates easily with existing sensors, network infrastructure and end-user applications.

Figure 2.1 displays the functional overview of the BigData IoT platform. The topmost block represents the Cloud platform, the middle one is the network connectivity while the bottom one is the local deployment, including gateway and sensors. The following functional domains have been identified:

- The "Smart Applications" domain is the IoT application itself.
- The "Users Management" allows the management of the identification, roles and connections of users.
- The "Interoperable Service and Dynamic Workflow" domain allows application writing, deploying, hosting and execution.
- The "Processing and Analytic Engine", provides services of stream processing and data analytics.
- The "Network communication" domain provides the IoT connectivity.
- The "Embedded software" and Hardware domains represent the IoT gateway and sensors themselves.

2.3 Data Analytics for the IoT

The amount of IoT data coming from real-world smart objects with sensing, actuating, computing and communication capabilities is exploding. The sensors and devices are more and more deployed, within more applications and across industries. This section first explores the characteristics of this data. It then presents several data analytics techniques able process this data.

2.3.1 Characteristics of IoT Generated Data

The volume and quality of the data generated by IoT devices is very different from the traditional transaction-oriented business data. Coming from millions of sensors and sensor-enabled devices, IoT data is more dynamic,

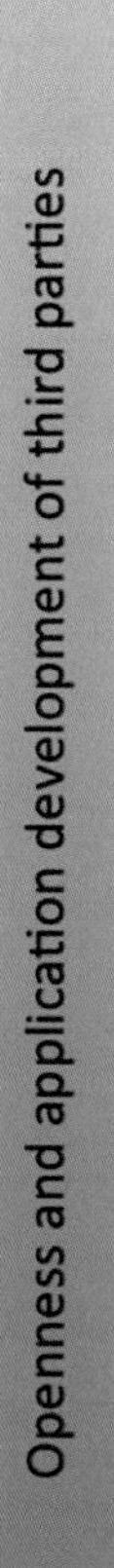

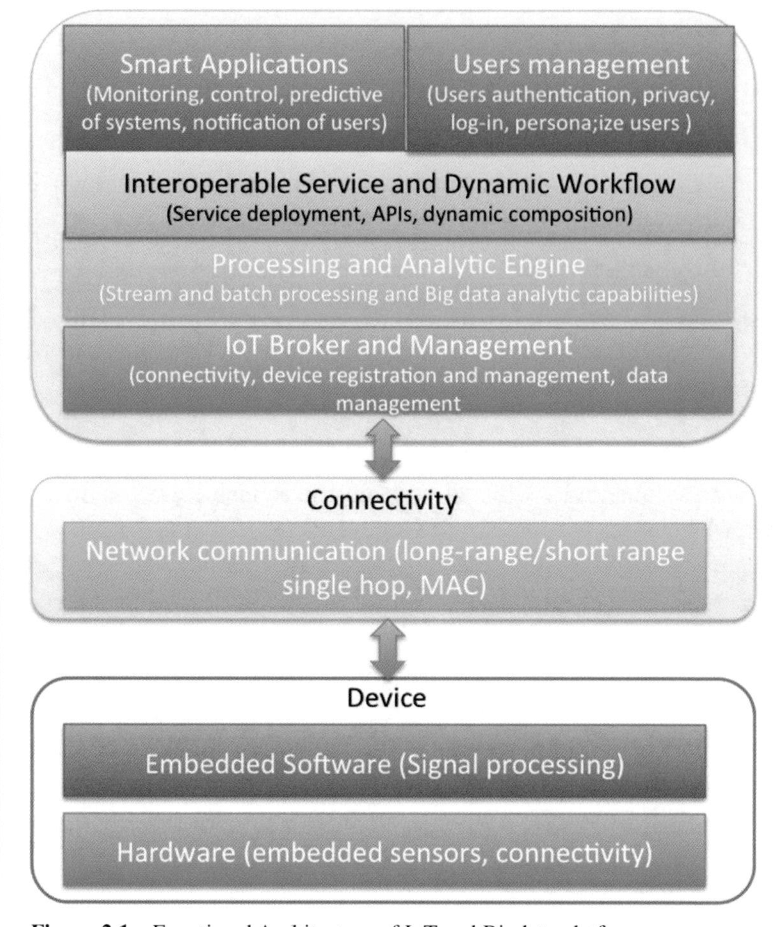

Figure 2.1 Functional Architecture of IoT and Bigdata platform.

heterogeneous, imperfect, unprocessed, unstructured and real-time than typical business data. It demands more sophisticated, IoT-specific analytics to make it meaningful.

As illustrated in Figure 2.2, the BigData is defined by 4 "Vs", which are Volume, Velocity, Variety and Veracity. The first V is for a large volume of data, not gigabytes but rather thousands of terabytes. The second V is referencing data streams and real-time processing. The third V is referencing the heterogeneity of the data: structure and unstructured, diverse data models, query language, and data sources. The fourth V is defining the data uncertainty,

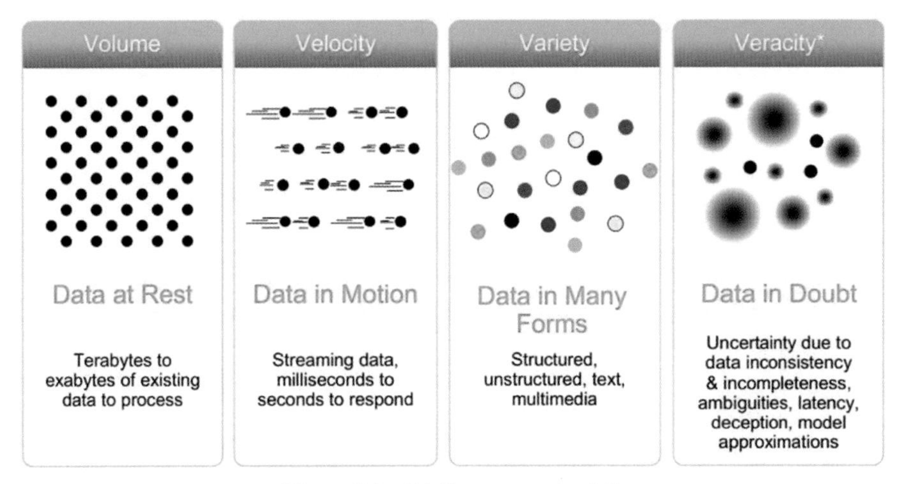

Figure 2.2 BigData properties [4].

which can be due to data inconsistency, incompleteness, ambiguities, latency and lack of precise model.

The IoT faces all 4 Vs of the BigData challenges. However the velocity is the main challenge: we need to process in real-time the data coming from IoT devices. For example, medical wearable such as Electro Cardio Graphic sensors produce up to 1000 events per second, which is a challenge for real-time processing. The volume of data is another important challenge. For example General Electric gathers each day 50 million pieces of data from 10 million sensors. A wearable sensor produces about 55 million data points per day. In addition, IoT also faces verity and veracity BigData challenges.

2.3.2 Data Analytic Techniques and Technologies

A cloud-based IoT analytics platform provides IoT-specific analytics that reduce the time, cost and required expertise to develop analytics-rich, vertical IoT applications. Platform's IoT-specific analytics uncover insights, create new information, monitor complex environments, make accurate predictions, and optimize business processes and operations. The applications of the IoT BigData Platform can be classified into four main categories i) deep understanding and insight knowledge ii) Real time actionable insight iii) Performance optimization and iv) proactive and predictive applications.

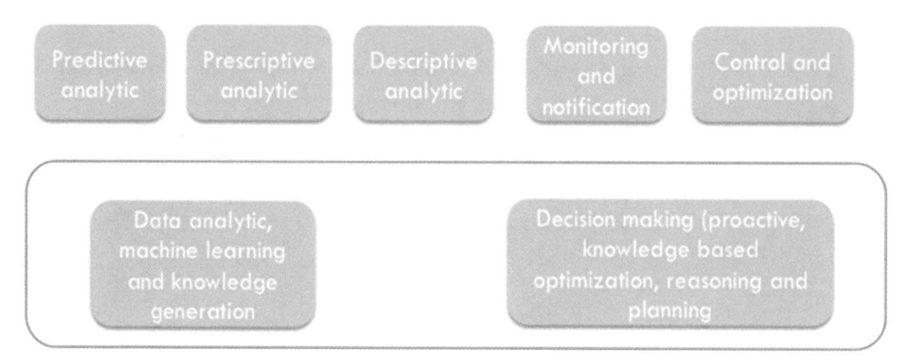

Figure 2.3 IoT BigData applications.

In the following we provide various technologies allowing building such an IoT analytics platform.

Batch Processing

Batch processing supposes that the data to be treated is present in a database. The most widely used tool for the case is **Hadoop MapReduce**. MapReduce is a programming model and Hadoop an implementation, allowing processing large data sets with a parallel, distributed algorithm on a cluster. It can run on inexpensive hardware, lowering the cost of a computing cluster. The latest version of MapReduce is YARN, called also MapReduce 2.0. **Pig** provides a higher level of programming, on top of MapReduce. It has its own language, PigLatin, similar to SQL. Pig Engine parses, optimizes and automatically executes PigLatin scripts as a series of MapReduce jobs on a Hadoop cluster. Apache **Spark** is a fast and general-purpose cluster computing system. It provides high-level APIs in Java, Scala, Python and R, and an optimized engine that supports general execution graphs. It can be up to a hundred times faster than MapReduce with its capacity to work in-memory, allowing keeping large working datasets in memory between jobs, reducing considerably the latency. It supports batch and stream processing.

Stream Processing

Stream processing is a computer programming paradigm, equivalent to dataflow programming and reactive programming, which allows some applications to more easily exploit a limited form of parallel processing. **Flink** is a streaming dataflow engine that provides data distribution, communication and fault tolerance. It has almost no latency as the data are streamed in real-time (row by row). It runs on YARN and works with its own extended version of MapReduce.

Machine Learning

Machine learning is the field of study that gives computers the ability to learn without being explicitly programmed. It is especially useful in the context of IoT when some properties of the data collected need to be discovered automatically. Apache Spark comes with its own machine learning library, called **MLib**. It consists of common learning algorithms and utilities, including classification, regression, clustering, collaborative filtering, dimensionality reduction. Algorithms can be grouped in 3 domains of actions: Classification, association and clustering. To choose an algorithm, different parameters must be considered: scalability, robustness, transparency and proportionality. **KNIME** is an analytic platform that allows the user to process the data in a user-friendly graphical interface. It allows training of models and evaluation of different machine learning algorithms rapidly. If the workflow is already deployed on Hadoop, **Mahout,** a machine learning library can be used. Spark also has his own machine learning library called MLib.

H20 is a software dedicated to machine-learning, which can be deployed on Hadoop and Spark. It has an easy to use Web interface, which makes possible to combine BigData analytics easily with machine learning algorithm to train models.

Data Visualisation

Freeboard offers simple dashboards, which are readily useable sets of widgets able to display data. There is a direct Orion Fiware connector. Freeboard offers a REST API allowing controlling of the displays. **Tableau Public** is a free service that lets anyone publish interactive data to the web. Once on the web, anyone can interact with the data, download it, or create their own visualizations of it. No programming skills are required. Tableau allows the upload of analysed data from .csv format, for instance. The visualisation tool is very powerful and allows a deep exploration the data. **Kibana** is an open source analytics and visualization platform designed to work with Elasticsearch. Kibana allows searching, viewing, and interacting with data stored in Elasticsearch indices. It can perform advanced data analysis and visualize data in a variety of charts, tables, and maps. **Elasticsearch** is a highly scalable open-source full-text search and analytics engine. It allows to store, search, and analyze big volumes of data quickly and in near real time.

It is generally used as the underlying engine/technology that powers applications that have complex search features and requirements. It provides a distributed, multitenant-capable full-text search engine with an HTTP web interface and schema-free JSON documents. It is really designed for real-time analytics, most commonly used with Flink or Spark streaming.

2.4 Data Collection Using Low-power, Long-range Radios

Regarding the deployment of IoT devices in a large scale, it is still held back by technical challenges such as short communication distances. Using the traditional mobile telecommunications infrastructure is still very expensive (e.g., GSM/GPRS, 3G/4G) and not energy efficient for autonomous devices that must run on battery for months. During the last decade, low-power but short-range radio such as IEEE 802.15.4 radio have been considered by the WSN community with multi-hop routing to overcome the limited transmission range. While such short-range communications can eventually be realized on smart cities infrastructures where high node density with powering facility can be achieved, it can hardly be generalized for the large majority of surveillance applications that need to be deployed in isolated or rural environments. Future 5G/LTE standards do have the IoT orientation but these technologies and standards are not ready yet while the demand is already high.

Recent so-called Low-Power Wide Area Networks (LPWAN) such as those based on Sigfox^{TM} or Semtech's LoRa^{TM} [1] technology definitely provide a better connectivity answer for IoT as several kilometers can be achieved without relay nodes to reach a central gateway or base station. Most of long-range technologies can achieve 20 km or higher range in LOS condition and about 2 km in urban NLOS [2]. With cost and network availability constraints, LoRa technology, which can be privately deployed in a given area without any operator, has a clear advantage over Sigfox which coverage is entirely operator-managed. These low-power, long-range radio technologies will definitely allow a huge amount of sensors to be installed in remote area, thus augmenting the amount of data to be treated in the IoT Cloud platform.

2.4.1 Architecture and Deployment

The deployment of LPWAN (both operator-based and privately-owned scenarios) is centred on gateways that usually have Internet connectivity as shown in Figure 2.4. Although direct communications between devices are possible, most of IoT applications follow the gateway-centric approach with mainly

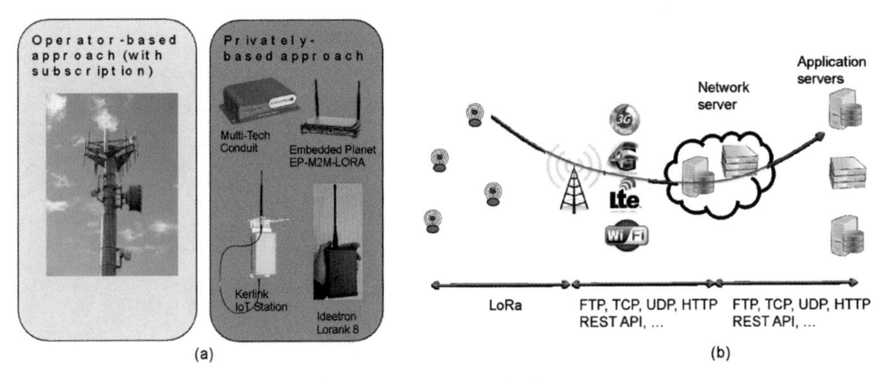

Figure 2.4 Gateway-centric deployment.

uplink traffic patterns. In this typical architecture data captured by end-devices are sent to a gateway which will push data to well identified network servers. Then application servers managed by end-users could retrieve data from the network server. If encryption is used for confidentiality, the application server can be the place where data could be decrypted and presented to end-users.

The LoRa Alliance has issued the LoRaWAN specification [3] in a tentative for standardization of public, large-scale LoRa LPWAN infrastructures featuring multi-gateways and full network/application servers' architecture as previously depicted in Figure 2.4. This specification also defines the set of common channels for communications, the packet format, Medium Access Control (MAC) commands that must be provided and 3 end-devices classes depending on communication requirements. This architecture can however be greatly simplified for small, ad-hoc deployment scenarios where the gateway can directly push data to some servers or IoT-specific cloud platforms if properly configured.

2.4.2 Low-cost LoRa Implementation

The implementation of the full LoRaWAN specification requires gateways to be able to listen on several channels and LoRa settings simultaneously. Commercial gateways therefore use advanced concentrators chips capable of scanning up to 8 different channels: the SX1301 concentrator is typically used instead of the SX127x chip which is designed for end-devices. Commercial gateways cost several hundredth euros with the cost of the SX1301-capable board alone to be more than a hundred euro.

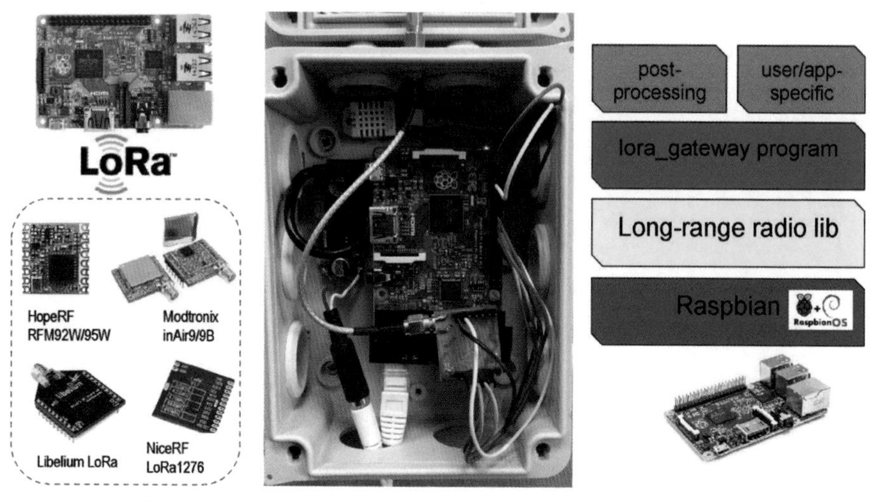

Figure 2.5 Low cost gateway from off-the-sheves components.

For many adhoc applications, it is however more important to keep the cost of the gateway low and to target small to medium size deployment scenario for various specific use cases instead of the large-scale, multi-purpose deployment scenarios defined by LoRaWAN. Note that even though several gateways can be deployed to serve several channel settings if needed. In many cases, this solution presents the advantage of being more optimal in terms of cost as incremental deployment can be realized and also offer a higher level of redundancy that can be an important requirement in developing countries for instance.

Our LoRa gateway could be qualified as "single connection" as it is built around an SX1272/76, much like an end-device would be. The cost argument, along with the statement that too integrated components are difficult to repair and/or replace in the context of developing countries, also made the "off-the-shelves" design orientation an obvious choice. Our low-cost gateway is based on a Raspberry PI (1B/1B+/2B/3B) which is both a low-cost (less than 30 euro) and a reliable embedded Linux platform. Our long-range communication library supports a large number of LoRa radio modules (most of SPI-based radio modules). The total cost of the gateway can be as low as 45 euro.

Together with the "off-the-shelves" component approach, the software stack is completely open-source: (a) the Raspberry runs a regular Raspian distribution; (b) our long range communication library is based on the SX1272 library written initially by Libelium and (c) the lora_gateway program is kept as simple as possible. We improved the original SX1272 library in various ways to provide enhanced radio channel access (CSMA-like with SIFS/DIFS) and support for both SX1272 and SX1276 chips. We believe the whole architecture and software stack are both robust and simple for either "out-of-the-box" utilization or quick customization by third parties.

We tested the gateway in various conditions for several months with a DHT22 sensor to monitor the temperature and humidity level inside the case. Our tests show that the low-cost gateway can be deployed in outdoor conditions with the appropriate casing. Although the gateway should be powered, its consumption is about 350 mA for an RPIv3B with both WiFi and Bluetooth activated.

2.5 WAZIUP Software Platform

The WAZIUP project, namely the Open Innovation Platform for IoT-BigData in Sub-Saharan Africa is a collaborative research project using cutting edge technology applying IoT and BigData to improve the working conditions in the rural ecosystem of Sub-Saharan Africa. First, WAZIUP operates by involving farmers and breeders in order to define the platform specifications in focused validation cases. Second, while tackling challenges which are specific to the rural ecosystem, it also engages the flourishing ICT ecosystem in those countries by fostering new tools and good practices, entrepreneurship and start-ups. Aimed at boosting the ICT sector, WAZIUP proposes solutions aiming at long term sustainability.

The consortium of WAZIUP involves 7 partners from 4 African countries and partners from 5 EU countries combining business developers, technology experts and local Africa companies operating in agriculture and ICT. The project involves also regional hubs with the aim to promote the results to the widest base in the region.

2.5.1 Main Challenges

The WAZIUP Cloud platform needs to face a number of challenges. Those challenges are related to the specific environment in which the platform will be deployed, and the need of its end users. First of all, we identified that farmers in Sub-Saharan Africa are lacking data on culture status. For instance, parameters such as potassium and nitrogen levels are very useful for precision farming. Secondly, farmers are lacking actionable information on the condition of the farm. This actionable information can be displayed in the form of alerts, forecasts and recommendations. An example of such a service is a recommendation on the water levels needed for irrigation, taking into accounts the weather forecasts. On a larger scale, governments and institutions are lacking information and statistics on their territory. An example is geographical statistics on the spreading of a disease in a country.

On a more technical level, we noticed that most rural African users have mobile phones, but not always smart phones. Furthermore, 3G is not always present in rural areas. Internet and grid connection can also be intermittent. Lastly, a huge challenge that the WAZIUP platform should address is the cost of IoT devices, application development and application hosting.

2.5.2 PaaS for IoT

As introduced before, PaaS framework will compile an application from its source code, and then deploy it inside lightweight virtual machines, or containers. This compilation and deployment is done with the help of a file called the manifest, which allows the developer to describe the configuration and resource needs for his application. The manifest file will also describe the services that the application requires and that the platform will need to provision.

The idea of WAZIUP is to extend the paradigm of the PaaS to IoT. Indeed, developing an IoT BigData application is a complex task. A lot of services need to be installed and configured, such as databases and complex event processing engines. Furthermore, it requires an advanced knowledge and skills in programing of embedded devices, of data stream processors, of advanced data analytics, and finally of GUIs and user interactions. We propose to abstract those skills using the PaaS paradigm.

Figure 2.6 shows the PaaS deployment in WAZIUP. Traditional PaaS environment are usually installed on top of IaaS (in blue in the picture). The blue boxes are physical servers, respectively the Cloud Controller and one Compute node. The PaaS environment is then installed inside the IaaS V Ms, in green in the picture. We use Cloud Foundry as a PaaS framework. It comes with a certain number of build packs, which and programming languages compilers and run time environments. It also provides a certain number of preinstalled services such as MongoBD or Apache Tomcat. The manifest file, showed on the right hand side, provide a high-level language that allows describing which services to instantiate. We propose to extend this language to IoT and BigData services:

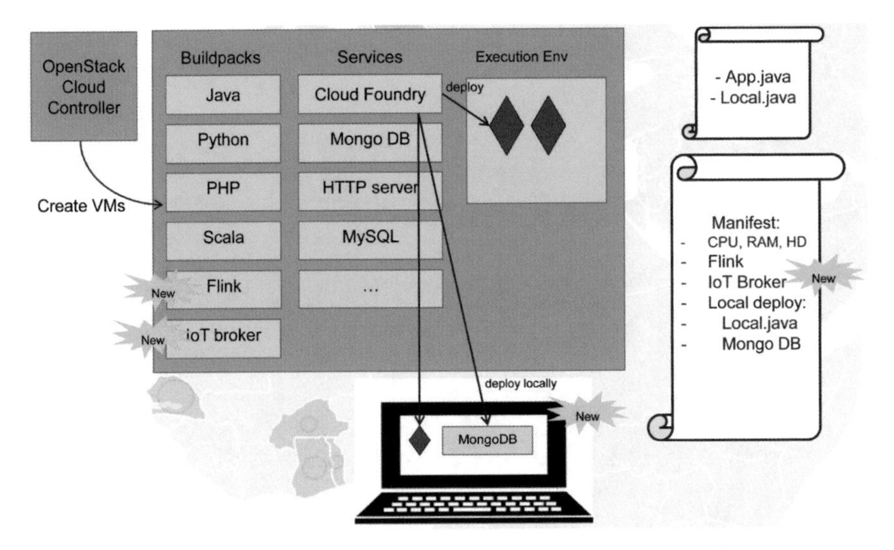

Figure 2.6 PaaS deployment extended for IoT in WAZIUP.

- Data stream and message broker
- CEP engines
- Batch processing engines
- Data visualization engines

Furthermore, we propose to include in the manifest a description of the IoT sensors that are required by the application. This query includes data such as the sensor type, location and owner. The manifest also includes the configuration of the sensors. The application will then be deployed both in the global Cloud and in the local Cloud.

2.5.3 Architecture

Figure 2.7 presents the full WAZIUP architecture. There are 4 silos (from left to right): Application development, BigData platform, IoT platform, Sensors and data sources. The first silo involves the development of the application itself. A rapid application development (RAD) tool can be used, such as Node-Red. The user provides the code source of the application, together with the manifest. As a reminder, the manifest describes the requirements of the application in terms of:

- Computation needs (i.e. RAM, CPU, disk).
- Reference to data sources (i.e. sensors, internet – sources . . .).
- BigData engines needed (i.e. Flink, Hadoop . . .).

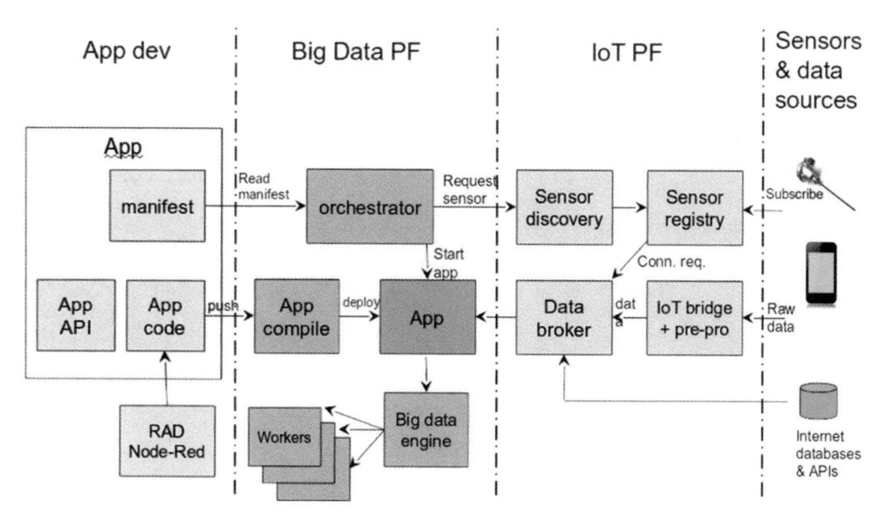

Figure 2.7 WAZIUP architecture.

- Configuration of sensors (i.e. sampling rate).
- Local and global application deployment.

The application source code, together with the manifest, is pushed to the WAZIUP Cloud platform by the user. The orchestrator component will read the manifest and trigger the compilation of the application. It will then deploy the application in the Cloud execution environment. It will also instantiate the services needed by the application, as described in the manifest. The last task of the orchestrator is to request the sensor and data sources connections from the IoT components of the architecture. The sensor discovery module will be in charge of retrieving a list of sensors that matches the manifest description.

On the left side of the diagram, the sensor owners can register their sensors with the platform. External data sources such as Internet APIs can also be connected directly to the data broker. The sensors selected for each application will deliver their data to the data broker, through the IoT bridge and preprocessor. This last component is in charge of managing the connection and configuration of the sensors. Furthermore, it will contain the routines for pre-processing the data transmitted, such as cleaning, extrapolating, aggregating and averaging data points.

2.5.4 Deployment

WAZIUP will be deployed and accessed in an African context, where internet access is sometime scarce. WAZIUP therefore has a very strong constraint regarding low internet connectivity. To fulfil this requirement, we propose a Cloud structure in two parts: the *global Cloud* and the *local Cloud*. The global Cloud corresponds to the Cloud in the traditional sense. The local Cloud corresponds to the gateway and an optional connected computer. The idea of WAZIUP is to extend the PaaS concept to the local Cloud.

A typical WAZIUP deployment is illustrated in Figure 2.8. On the left hand side of the picture, the application is designed by the developer, together with the manifest file. It is pushed on the WAZIUP Cloud platform. The orchestrator then takes care of compiling and deploying the application in the various Cloud execution environments. Furthermore, the orchestrator drives the instantiation of the services in the Cloud, according to the manifest. The manifest is also describing which part of the application need to be installed locally, together with corresponding services. The local application can then connect to the gateway and collect data from the sensors.

2.6 iKaaS Software Platform

The iKaaS platform combines ubiquitous and heterogeneous sensing, BigData and cloud computing technologies in a platform enabling the Internet of Things process consisting of continuous iterations on data ingestion, data storage, analytics, knowledge generation and knowledge sharing phases, as foundation service provision.

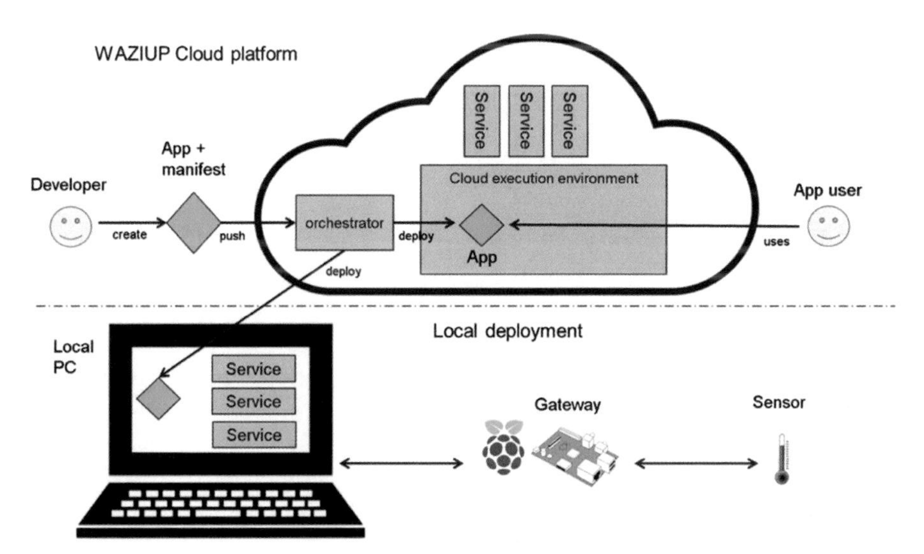

Figure 2.8 WAZIUP local and global deployment.

The iKaaS platform consists of two distinct Cloud ecosystems: the *Local Cloud* and the *Global Cloud*. More specifically:

- A Local Cloud provides requested services to users in a limited geographical area. It offers additional processing and storage capability to services. It is created on-demand, and comprises appropriate computing, storage and networking capabilities.
- The Global Cloud is seen in the "traditional" sense, as a construct with on-demand and elastic processing power and storage capability. It is a "backbone infrastructure", which increases the business opportunities for service providers, the ubiquity, reliability and scalability of service provision.

Local Clouds can involve an arbitrarily large number of nodes (sensors, actuators, smartphones, etc.). The aggregation of resources comprises sufficient processing power and storage space. The goal is to serve users of a certain area. In this respect, a Local Cloud is a virtualised processing, storage and networking environment, which comprises IoT devices in the vicinity of the users. Users will exploit the various services composed of the Local Cloud's devices' capabilities. For example, a sensor and its gateway equipped with the iKaaS platform.

The Global Cloud allows IoT service providers to exploit larger scale services without owning actual IoT infrastructure.

The iKaaS Cloud ecosystem will encompass the following essential functionality:

- Consolidated service-logic, resource descriptions and registries will be parts of the Global Cloud. These will enable the reuse of services. Practically, a set of registries will be developed and pooling of service logic and resources will be enabled.
- Autonomic service management will be part, firstly, of the Global Cloud, and, then, in the Local Clouds. This functionality will be in charge of (i) dynamically understanding the requirements, decomposing the service (finding the components that are needed); (ii) finding the best service configuration and migration (service component deployment) pattern; (iii) during the service execution, reconfiguring the service, i.e., conducting dynamic additions, cessations, substitutions of components.
- Distributed data storage and processing is anticipated for the structure of global and local clouds. This means capabilities for efficiently

communicating, processing and storing massive amounts of, quickly-emerging, versatile data (i.e., "BigData"), produced by a huge number of diverse IoT devices. Another important capability will be the derivation of information and knowledge (e.g., on device behaviour, service provision, user aspects, etc.), while ensuring security and privacy, which are top concerns.

- Knowledge as a service (KaaS) will be primarily part of the Global Cloud. This area will cover: (i) device behaviour aspects; (ii) the way services have been provided (e.g., through which IoT resources) and the respective quality levels; (iii) user preferences.

As can be seen the iKaaS functionality will determine the optimal way to offer a service. For instance service components may need to be migrated as close as possible to the required (IoT) data sources. IoT services may need generic service support functionality that is offered within the Cloud, and, at the same time, they do rely on local information (e.g., streams of data collected by sensors in a given geographic area), therefore, the migration of components close to the data sources will help in the reduction of the data traffic.

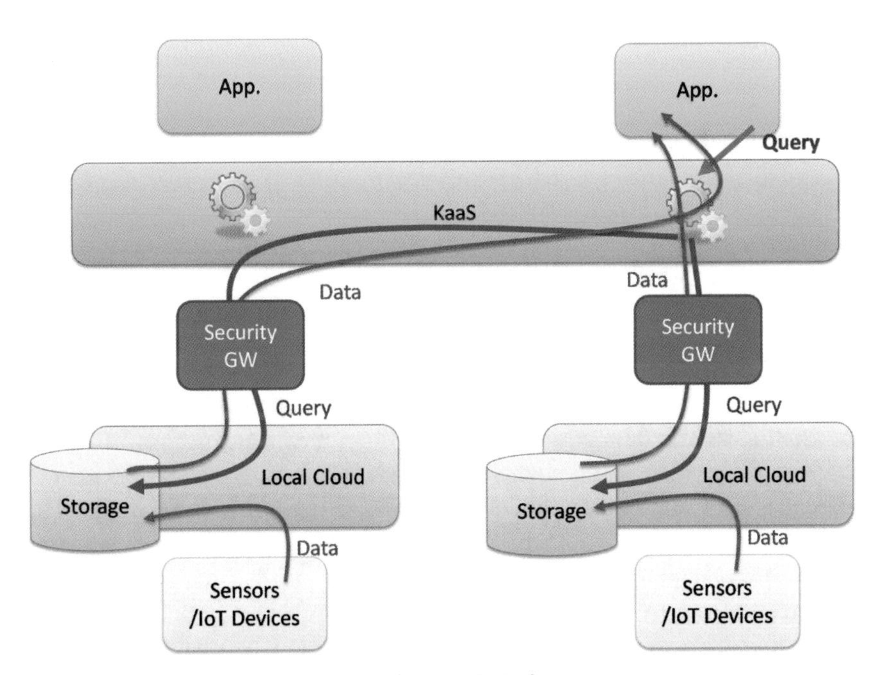

Figure 2.9 iKaaS platform.

2.6.1 Service Orchestration and Resources Provisioning

The platform offers mechanisms that autonomously analyse application requirements, user preferences and Cloud resources and accordingly decide upon the most appropriate deployment of services. The most appropriate deployment must achieve the best balance between system performance, quality of service and cost. In this context, services may be decomposed into smaller components, based on the current situation and information on data sources, in order to be migrated and executed in a "Local Cloud", near the data sources, following the Hadoop maxim that "*Moving Computation is Cheaper than Moving Data*". Alternatively, services may be deployed and executed in the Global Cloud. Furthermore, this mechanism will facilitate the notion of "Everything as a Service", and attached gateway to host and process services on-demand, by means of service migration instead of being limited to predefined services. The local IoT Gateway may act as part of a "Local Cloud" on an on-demand basis in coordination with the Global Cloud, provided that the Local Cloud has sufficient resources to process and execute the service.

The platform uses a model that allows the service to be analysed and decomposed into a certain number of sub-components according to application requirements, user preferences including privacy constraints, policies, system state and data sources location. The service sub-components are then migrated to either Local Clouds, to be computed near the data sources (e.g., sensors) or into the Global Cloud, to take advantage of the extensive computing power and storage available. The optimal distribution is decided with the aim of achieving the best balance between overall system performance (network traffic, computing load), quality of services (prompt and accurate delivery of service result) and service costs.

2.6.2 Advanced Data Processing and Analytics

Information stream processing algorithms and mechanisms offer on-the-fly analysis of volatile data coming from the distributed sensing infrastructure. In addition, the platform includes off-line BigData analytics over persistent data capable of uncovering hidden patterns and unknown correlations. This will allow feeding with contents the envisaged knowledge service platform.

The iKaaS platform includes the analysis of the requirements and challenges posed by those information stream processing and knowledge acquisition scenarios to the provision of a set of IoT and BigData services over cloud and network infrastructures.

The provision of those services comprises processing capabilities, covering the knowledge acquisition lifecycle. This lifecycle goes from the aggregation of heterogeneous data, through information stream processing services and visualization services, to the derivation of knowledge and experience. Special attention will be paid to the consolidation of existing approaches and to the design of complementary solutions able to address the technological challenges:

- Information Stream Processing, information extraction and visualization, mechanisms enabling the usage of smart virtual objects as a multi-cloud cloud based resource.
- Distributed and scalable storage mechanisms for smart virtual objects that supports service decomposition, migration and corresponding resource allocation aspects within the iKaaS local and Global Cloud environments.
- Analytics engines and mechanism for assessment and processing of data over a large number of smart objects. The objective is to derive reliable information and to provide knowledge that can be provided as a service to facilitate situation aware applications.

Given that processing and storage may take place in either the Global or the Local Cloud or both, in support of real-time autonomic and flexible service execution, the mechanisms defined in this task shall support flexible and fast discovery of smart virtual objects and allocation of data sources so that efficient and cost-effective service and resource migration can be realized.

Hence, the platform offers the mechanisms and techniques for handling smart objects and processing of their data to satisfy real-time service execution requirements in Cloud environments and also to derive useful "contextual" information and knowledge to serve cost-effective, low latency resources migration and allocation needs.

The scalable and distributed storage mechanism for smart virtual objects and aggregated and anonymised data will also need to be managed dynamically in order to deal with the large number of data sources.

2.6.3 Service Composition and Decomposition

Principle of Service Composition and Decomposition

IoT and BigData applications are complex large-scale applications, including a combination of multiple sources, functionalities and composed by many small functional services across multiple sectors/domains. For example, an active and healthy living of ageing people application includes many small services like monitoring the blood pressure, monitoring the heart rate, weight monitoring, location awareness, smart lighting, utility metering, notification and reminders, etc., across health, well-being, security and home automation domains. Additionally, for IoT and BigData in a given application as the service is evolving, more and more services added to the applications/systems. Therefore, it is important to design the iKaaS services as small and autonomous as possible, with well-defined APIs to operate them individually.

iKaaS functional decomposition of an application/complex service (as defined in the previous sub-section) allows to achieve loose coupling and high cohesion of multiple services. Alternatively multiple simple services can be composed into complex services for the purposes of various applications. In Figure 2.10, the basic logic of service decomposition and composition are shown.

Functional decomposition of services gives the agility, flexibility, scalability of individual services to operate autonomously. Each of the simple services is running in its own process and communicating with lightweight mechanisms. The overall high-level service logics (e.g. software module) are decomposed to multiple service logics or software modules which can be delivered as independent runtime services. These services are built around business capabilities and are independently deployable by fully automated deployment machinery. There is a bare minimum of centralized management

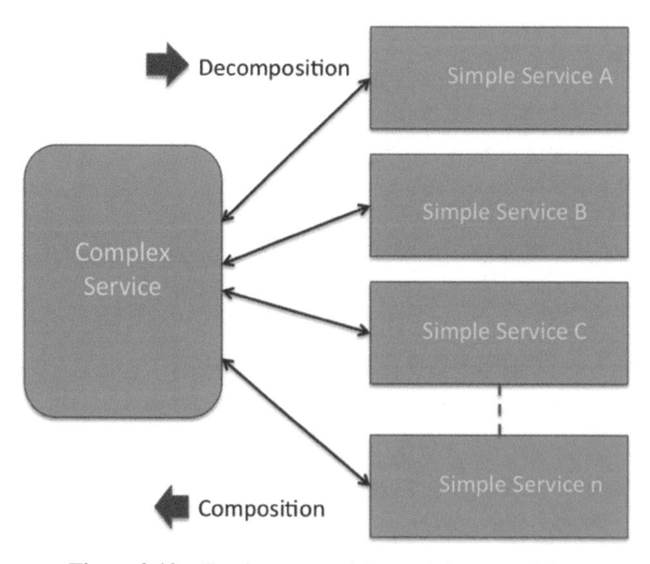

Figure 2.10 Service composition and decomposition.

of these services, which may be written in different programming languages and use different data storage technologies.

Pattern of Service Composition

The iKaaS service design pattern significantly impacts how the services will be composed and decomposed. One of the main concepts is to design the services as independent as possible. In the design pattern the service replication and reliability should also be considered. One individual complex service can be composed by multiple isolated end-users or system level services. The relationship between the services and datasets, whether each of the services is using its own dataset or sharing a single dataset with other services can vary. However each iKaaS simple service is associated with a relevant dataset in order to make the service fully independently designed and deployable.

At runtime, one iKaaS service may consist of multiple service instances. Each service instance is a runtime (e.g., Docker container). In order to be highly available, the containers are running on multiple Cloud VMs. In this case, the Service Manager acts as a load balancer that distributes requests across the service instances.

2.6.4 Migration and Portability in Multi-cloud Environment

Service migration is a concept used in cloud computing implementation models that ensures that an individual or organization can easily shift services

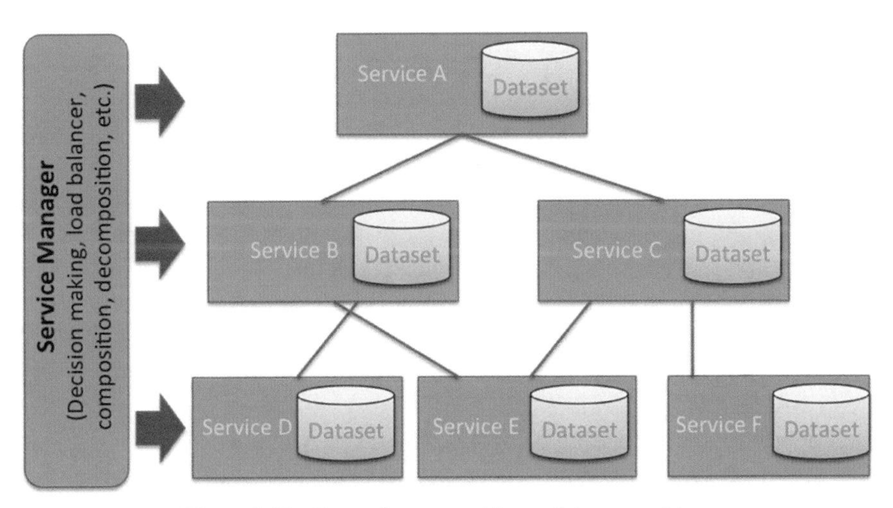

Figure 2.11 Patten for composition and decomposition.

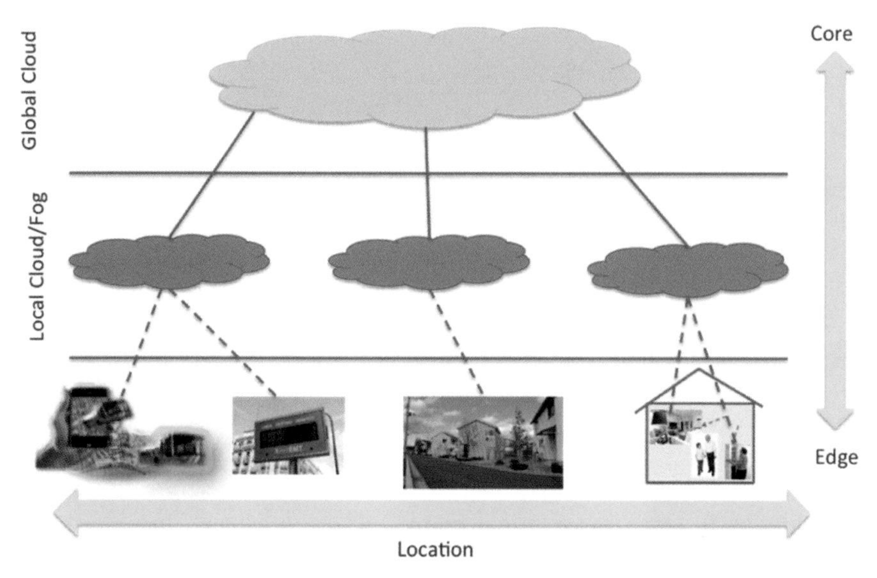

Figure 2.12 iKaaS distributed local and global cloud with service migration.

between different cloud vendors without encountering implementation, integration, and compatibility and interoperability issues. The concept is defined the process and framework by which these applications can be deployed on another cloud vendor or supported private cloud architecture.

iKaaS runtime puts services and all of its dependencies into a container which is portable among different platforms, desktop distributions and clouds. One can build, move and run distributed services with containers. By automating deployment inside containers, developers and system administrators can run the same application on laptops, virtual machines, bare-metal servers, and the cloud.

The concept of service migration is applicable in multi-cloud and distributed computing environment, where the processing capabilities are moved to near the data sources or a simple service is run near data sources. iKaaS is a fully distributed architecture, in which the overall platform functionalities and capabilities are distributed between local and global could. The concept of local could can be seen as an edge or fog computing, where pre-processing are done at the origin level.

Fog computing and computing near the data source provide a promising new approach to significantly reduce network operation cost by moving the computation or early pre-processing close to the data sources. A key challenge in such systems is to decide where and when services should be migrated with respect to users mobility, overall situation and environment context.

Edge/local computing can provide elastic resources to large scale data process system without suffering from the drawback of cloud, high latency. In cloud computing paradigm, event or data will be transmitted to the data centre inside core network and result will be sent back to end user after a series of processing. A federation of fog and cloud can handle the BigData acquisition, aggregation and pre-processing, reducing the data transportation and storage, balancing computation power on data processing. For example, in a large-scale environment monitoring system, local and regional data can be aggregated and mined at fog nodes providing timely feedback especially for emergency case such as toxic pollution alert. Detailed and thorough analysis as computational-intensive tasks can be scheduled in the cloud side.

2.6.5 Cost Function of Service Migration

One of the main challenges for the services migration is to define the strategy for the service migration. There is a tradeoff between the service migration cost and the transmission cost (such as communication delay and network overhead) between the user and the cloud. It is challenging to find the optimal decision also because of the uncertainty in user mobility as well as possible non-linearity of the migration and transmission costs. The service migration offers the benefits of reduction in networks overhead and latency over changing the location of the users. It is often challenging to make the optimal decision in an optimal manager, which can optimize the cost functions based on the situation and user's preferences.

iKaaS will propose a framework for dynamic, cost-minimizing migration of distribution services into a hybrid cloud infrastructure that spans geographically distributed data centers. We will propose an algorithm which optimally places services in different clouds to minimize overall operational cost over time, subject to service response time constraints. The framework will be designed based on the Markov-Decision-Process (MDP), to study service migration in iKaaS Cloud environment.

2.6.6 Dynamic Selection of Devices in Multi-cloud Environment

The end-devices are expected to play a key role in iKaaS not only for they data they can provide for the optimization of the iKaaS provided services, but also for the data they can provide with respect to end-device suitability and social relationships.

That is because in the case of end-device suitability, end-devices can be viewed as the end-points of an end-to-end service chain over a multi-cloud infrastructure. As such, this device suitability identification can be seen as "fixing" the end-points of the service provisioning chains, allowing as such (once the end-points are fixed) to then "fix" the location/placement of service provisioning functions in a way that optimizes both the service and the overall cloud performance. For example, if a device is identified as suitable, which is attached at a certain local cloud, then it would make sense for the other service functionalities needed to be instantiated at that local cloud so as to be close to the data source.

Some of the key factors that can be taken under consideration when defining device suitability are:

- Location/mobility pattern of a device; so as to define where the owner of the device is or is predicted to be so as to appropriately instantiate service functions close to the corresponding locations.
- Battery levels and evolution of battery levels; so as to be able to deduce whether a device can be relied upon for providing/receiving data, therefore corresponding service functions will need to be appropriately instantiated.
- Availability of sensors; how often a user has its device sensors exposed and is ready to be a potential match for inclusion in a service delivery chain.
- User away and reaction times; to make sure the user carries the device with them and is able to see an alert on time and react to it.
- Data quality: The quality of users inputs, without false-positives or misleading measurements.

All these factors will be further and better thought of and appropriate knowledge building mechanisms based on the nature and granularity of data will be considered. The scope is to eventually produce and store the knowledge about device suitability so that functionalities that decide on the placement of service functionalities in an end-to-end delivery chain, can take this into account when performing their joint service and cloud platform optimization processes.

Acknowledgement

This work has been produced in the context of the H2020 WAZIUP as well as H2020 iKaaS projects. The WAZIUP/iKaaS project consortium would like to acknowledge that the research leading to these results has received funding from the European Union's H2020 Research and Innovation Program (H2020-ICT-2015/H2020-ICT-2015).

References

[1] Semtech, "SX1276/77/78/79 – 137 MHz to 1020 MHz Low Power Long Range Transceiver. rev.4-03/2015," 2015.

[2] S. Jeff McKeown, "LoRa – a communications solution for emerging LPWAN, LPHAN and industrial sensing & IoT applications. http://cwbackoffice.co.uk/docs/jeff~20mckeown.pdf," accessed 13/01/2016.

[3] LoRa Alliance, "LoRaWAN specification, v1.01," Oct. 2015.

[4] Anders Quitzau, "Transforming Energy and Utilities through Big Data & Analytics" Big Data & Analytic, IBM, http://www.slideshare.net/Anders QuitzauIbm/big-data-analyticsin-energy-utilities

[5] www.waziup.eu

[6] www.ikaas.com

3

Searching the Internet of Things

Richard McCreadie[1], Dyaa Albakour[2], Jarana Manotumruksa[1], Craig Macdonald[1] and Iadh Ounis[1]

[1]University of Glasgow, UK
[2]Signal Media, UK

3.1 Introduction

Despite the huge success of Web search engines, searching the Web is far from being a solved problem (e.g. see [64] by Yahoo! Search). However, the information needs of the searchers are increasingly 'time-sensitive' – about events happening now – and/or 'local' – where the user's location has some geographical bearing on the content that is relevant to their information need(s). For instance, events that are happening now or recently may have an impact upon the searching behaviour of users. Indeed, a search engine can detect a power cut in New York within seconds, based on the querying behaviour of mobile and nearby users [30]. However, while a Web search engine can retrieve many forms of online information, it can only sense real-world events through their impact on the online world (e.g. news stories, tweets, increased query volume).

In this chapter, we describe how real-world information needs can be better addressed by search engines through harnessing sensing infrastructures, including those from the Internet of Things (IoT). Indeed, the introduction of IoT sensors within the search engine provides more responsive/timely information than existing evidence sources, such as Web or social streams, as illustrated in Figure 3.1. For instance, by considering IoT sensor outputs such as real-time rain levels, a search engine can produce a more customised answer to queries such as "what is the current weather at JFK airport?". Furthermore, information needs such as "what is happenning near me?" (local event retrieval) can be better answered by fusing social media trend data (also known as social sensing) with physical sensor observations. For example, a

Figure 3.1 Data sources available to an IoT-connected search engine.

party in the town square might be detected through a combination of people posting about it on Twitter, with crowd density analysis over CCTV camera feeds from the square. It is also possible to provide recommendations (e.g. for tourists) about points-of-interest to visit, that are appropriate to particular personalised interests of the users and their current contextual situation (time, location, travelling with friends, etc.). In doing so, search engines can help users satisfy new forms of information needs needs centred on real-world events.

In the following sections, we first provide details of search infrastructure technologies suitable for obtaining and indexing observations from a plethora of diverse IoT-connected sensors (Section 3.2); Later, we show how physical sensor information and socially-sensed information – combined with such technologies – can be adapted for tasks such as local event retrieval (Section 3.3), event topic identification (Section 3.4) and venue recommendations (Section 3.5). We conclude this chapter by discussing the outlook for the field and some interesting future directions and applications for IoT technologies in the search domain (Section 3.6).

3.2 A Search Architecture for Social and Physical Sensors

To achieve effective and efficient search over sensor data streams, it is important to have a suitable search architecture. Early exploration into the sensor space focused on the development of tools and techniques for searching sensor data using classical information retrieval techniques and architectures [17, 28]. These approaches exploit sensor ontologies [46] in order to decouple user queries from the low-level details of the underlying sensors. For instance, they might map a rain gauging data stream to particular weather-related queries, such that current rain data can be displayed when a user enters one of those queries. However, these ontologies are quite brittle in the face of changes in the user's query semantics and need to be hand-crafted for each domain/sensor stream. Hence, they cannot provide effective search over the arbitrary large and diverse sources of multimedia data derived from both physical and social sensors. Furthermore, sensor integration is only one of the components that are needed when building an IoT-enabled search engine. It is also critical to have effective and efficient indexing and retrieval processes over the sensor data, as well as have the ability to leverage the new search capabilities to build applications beyond the classic 'search bar'.

More recently, the SMART (Search engine for MultimediA enviRonment generated contenT) project[1] developed a framework [4] that aims to solve these issues, by providing an infrastructure where multimedia sensing devices in the social and physical world can be easily integrated into a central search system. By doing so, each sensor can provide information about their environment (physical or social) and make it available in real-time for search. As one of the most modern IoT search platforms in use today, we summarise the components that comprise SMART in Section 3.2.1. We then discuss some of the key challenges when building and deploying an effective IoT search engine like SMART in Section 3.2.2

3.2.1 Search engine for MultimediA enviRonment generated contenT (SMART)

SMART [4] is a framework designed to enable multimedia IoT sensing devices, both social and physical, to be integrated into a real-time search system. The architecture of the SMART framework is comprised of three distinct layers, as illustrated in Figure 3.2. At the lowest level we have the sensing devices that provide the physical world data. The edge node represents the software layer that processes the raw sensor data to produce metadata about the environment, which is streamed in real-time to the search engine using an appropriate representation. Examples of processing algorithms can include crowd data analysis for video streams or speech recognition in audio streams. The search layer collects the metadata streams from the various edge nodes and indexes them in real-time using an efficient distributed index structure. It also employs an event detection and ranking retrieval model that uses features extracted over the metadata streams to satisfy the user's information need. For instance, as will be described later in our discussion on search applications in

[1] http://www.smartfp7.eu/

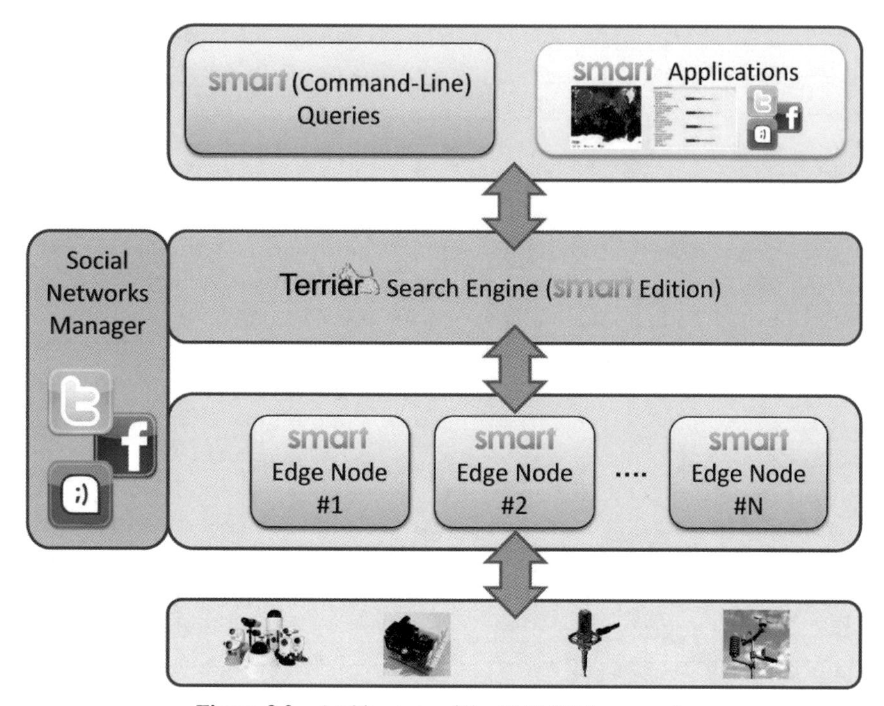

Figure 3.2 Architecture of the SMART framework.

Sections 3.3, and 3.4, the search service can be used to rank real-world events detected from the sensor streams that a user might be interested in attending. Queries can be either directly specified, or anticipated by the search layer using contextual information about the user, e.g. the user's location or their social profile. Finally the *application/visualisation layer* at the top offers reusable APIs to develop applications that can issue queries to the SMART engine and process or visualise the results. We further describe these three layers in more detail below.

Edge Node Layer

The edge node is the interface of SMART with the physical world. Each edge node can cover sensors from a single geographic area, e.g. a building block or a public square in the city centre. At each edge node, signal streams are processed to extract events/patterns that might be of value for answering one or more information needs. The signal streams can either be derived from physical sensors (e.g. audio/visual streams or environmental measurements), or from real-time Web crawling/social network streams. To achieve this, the

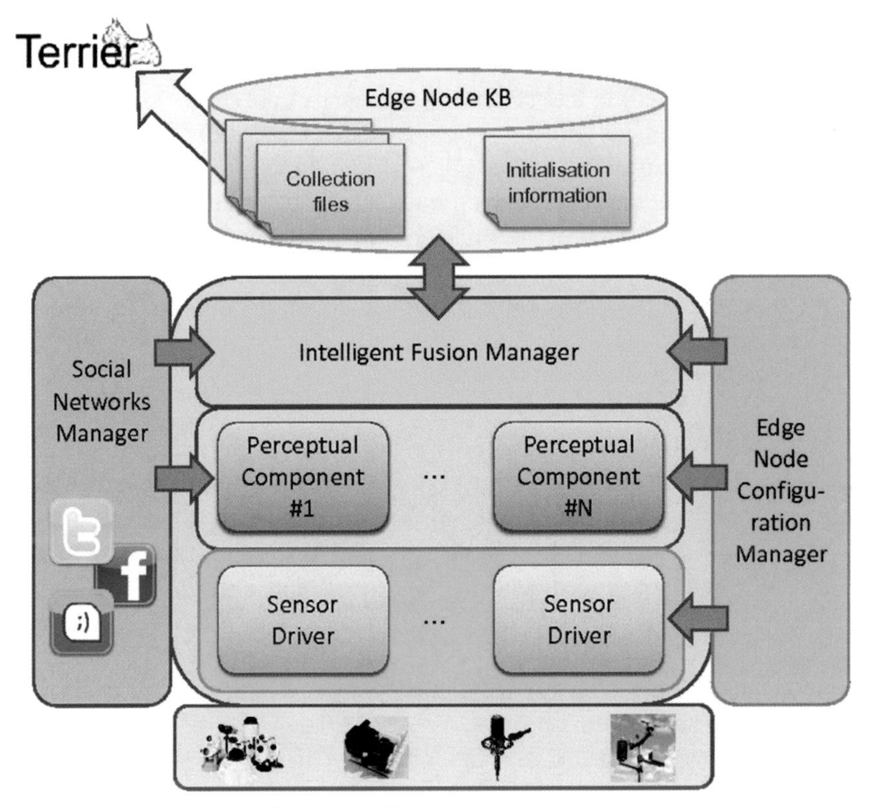

Figure 3.3 Edge node components.

design of the edge node is influenced by state-of-the-art IoT platforms and Linked Data techniques. The edge node architecture is shown in Figure 3.3. As we can see from Figure 3.3, at the lowest level of the edge node lies the sensors themselves. These sensors are interfaced with via sensor drivers, that allow for the connection to the sensor and the streaming ingestion of the raw sensor data into the edge node. The raw sensor data is then subject to processing by one or more perceptual components, which convert that data into a form that is more actionable. For instance, a perceptual component for an air quality sensor might take a CO2 reading and convert it to a label such as 'Normal' or 'High' based on external knowledge about what CO2 levels are acceptable. Next, the processed sensor outputs are sent to the Intelligent Fusion Manager of the edge node. This manager enables the reasoning over the outputs of different sensors within that edge node concurrently. For instance, for an edge node responsible for tracking a shopping street, with physical CCTV camera streams spaced along that street and a social sensor looking for posts geo-located within that street, the manager might merge crowd signals from the CCTV cameras with the volume of social media activity to predict the number of people currently shopping there. Finally, the output of the intelligent fusion manager and the perceptual components feed an edge node knowledge base, which stores the observations made over the sensors across time. For instance, continuing the shopping street example above, the knowledge base would store the population estimates for the street at different times of the day. The edge node knowledge base content is stored as series of collection files that can be indexed by the search layer.

Search Engine Layer

The SMART search layer indexes in real-time streams of collection files from edge nodes, along with other conventional streams (such as social network posts or Web documents). It is built using the Terrier[2] open source search engine [48] with enhanced real-time indexing and a scalable distributed architecture to handle the large amount of streams. The SMART search layer is comprised of 7 core components as illustrated in Figure 3.4. The Indexing Component is responsible for the representation, storage and organisation of the information streams provided, such that they are available for later retrieval. It ingests the streams of collection files from edge nodes and social/Web documents (via data feed connectors), and performs a real-time indexing of those streams into appropriate data structures that allow for efficient retrieval. Real-time indexing ensures that as soon as an item (such as a social media post or street density summary) arrives on one of the input streams, that item will be searchable immediately. The index is distributed across multiple index shards (machines) so as to cope with a potentially high number and volume of social sensor streams, ensuring the scalability of the overall system architecture. This is achieved through the use of the distributed stream processing platform Storm.[3] Storm is one of the new generation of distributed real-time computation platforms, which provides an easy means to distribute complex software topologies across multiple machines, while maintaining fault tolerance and low management overheads. In this case, the content indexing pipeline is represented as a series of processing nodes (known as 'bolts'), where each node/bolt can be replicated and distributed across a local

[2]http://terrier.org

[3]http://storm.apache.org/

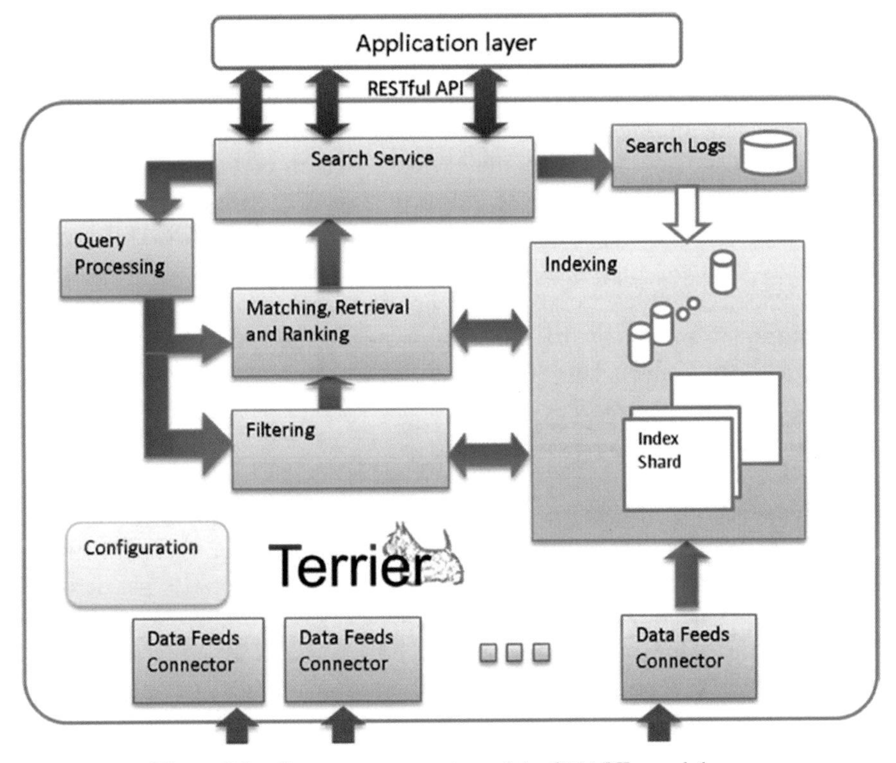

Figure 3.4 Components overview of the SMART search layer.

or cloud service machine cluster to achieve scalability. Next, the query processing component identifies the user information needs as specified explicitly by the user. Queries can be anticipated or expanded by observing past occurring patterns. The Matching, Retrieval and Ranking Component is responsible for matching explicit or implicit user queries against the index to *identify, rank* and *recommend* events/locations according to how they satisfy the users' information needs. This component relies on newly developed retrieval and recommendation models that can identify interesting "unusual" events across sensor (inc. social) metadata streams. The Filtering Component identifies in real-time events (or social network posts) as they happen that match a user's running query. This permits a user to be notified of new events that they will find interesting. This component handles queries after they have been submitted to the SMART search layer (as running queries) so that updates are streamed back to the higher level applications in real-time. The Search Logs Component maintains a recording of the search behaviour of the user population. The search behaviour includes the user interactions with the application such as the queries that have been issued in a user search session, which search results have been displayed and any documents clicked by the user. This implicit feedback obtained by monitoring the users' search behaviour can be fed back into the SMART search layer, for example, to improve the effectiveness of the search results or the recommendations. The Search Engine API Component provides an interface to the SMART search layer where the main functionalities (search and running queries) are defined and are made available to higher level applications or services. Finally, the Configuration Component offers a series of administrative functionalities, such as the setup of the data streams to use as input and the choice of the matching algorithms to deploy.

Application Layer

The top layer of the SMART platform contains the software applications that can deliver the real benefits of the framework to the end-user. The application layer mainly supports developers who want to create Web 2.0 services or smart phone applications that exploit the framework capabilities. For example, the application layer includes open source end-user web applications that offer user interfaces to issue queries explicitly, or implicitly using the user context, to the search engine API and receive in real-time up-to-date results (events). In addition, it includes open source mashups that use the search layer visualisation APIs to display newly-breaking events, such as real-time balloon pop-ups on a map.

3.2.2 Challenges in Building an IoT Search Engine

Importantly, there are a variety of challenges when implementing an IoT search architecture like SMART. First, data stream collection and processing algorithms are needed to provide a uniform means to interface with a wide array of sensor types and to perform processing on those sensors' output to make that output interpretable/useful to the search engine. For instance, a raw video feed cannot be directly used to answer a user's information need. However, processing that feed through crowd analysis software to get crowd density for a street might be useful to predict the number of people visiting the area. Furthermore, some types of sensor streams require pre-filtering to make them useful. For example, it might be advantageous to define a social sensor, by filtering down a wider stream of posts to only those from a particular geographical region [2]. Within SMART, functionality like this is performed by the perceptual components within each edge node. However, to incorporate the ever-growing range of IoT devices, new processing algorithms tailored to these devices will be needed.

Second, a common metadata model is needed to enable the processed sensor outputs to be mapped into a standardised metadata stream [12]. This is important, since often individual sensors only act as weak indicators of some higher level activity that the user might want to search for. For instance, if we want to detect live music in a city square, we might want to combine evidence from social sensors like discussions on Twitter, with physical evidence such as a locally captured audio or crowd density analysis from the square (c.f. Section 3.4). The use of a common metadata model can facilitate concurrent reasoning across multiple sensor streams by mapping lots of weak metadata signals from different sensors into the same format. For instance, SMART uses a model based on the OGC's Sensor Web Enablement standards [13] within the Intelligent Fusion Manager to achieve this.

Next, within the search engine itself, the efficient real-time indexing of the underlying metadata streams is critical. In particular, in an IoT environment, thousands of sensors can be feeding the search engine concurrently, and users expect the most up-to-date results. Hence, the search engine needs to be able to ingest high volume sensor streams in real-time while concurrently serving search requests over the most recent data. To achieve this, distributed stream processing platforms such as Storm[4] or Apache Spark[5] are used, as they allow for the low-latency processing of content in a distributed scale-out manner.

Finally, the types of queries and underlying information needs within the IoT search space are markedly different to those observed within a classical Web search domain. As a consequence, new retrieval models designed for these novel information needs are required. For instance, for an event search engine, a model that can effectively rank current (and possibly predict future) events based on criteria such as relevance, interestingness to the user or timeliness, are needed. Furthermore, in some applications, such as venue suggestions (that we will cover in detail later in Section 3.5), additional criteria needs to be considered, such as the user location (and hence distance to the event) and other contextual features such as the time of the day or the current weather. Current systems rely on state-of-the-art learning-to-rank techniques [39] to learn an effective combination of these diverse types of evidence when ranking.

---

[4]https://github.com/nathanmarz/storm/

[5]http://spark.apache.org/

In the remainder of this chapter we discuss three recent applications of the SMART framework that examined how to satisfy new user information needs using social and IoT sensing. In particular, we discuss social sensing with SMART for event retrieval in Section 3.3. Section 3.4 describes an application where IoT sensor streams were fused with social evidence for event topic identification. Finally, we discuss context sensitive venues-recommendation based on social sensing in Section 3.5.

3.3 Local Event Retrieval

It has been suggested that a large proportion of queries submitted to web search engines has a "local intent" and that these queries compose the majority of searches submitted from mobile phones [58]. Examples of information needs expressed by such queries include "what is happening near me?" or "finding restaurants in the Covent Garden district". The prevalence of such queries highlights the importance of building effective local search tools that serve this type of information need. In this section, we present an approach for local event retrieval, where we rely solely on social media as a *social sensor* to detect events in real-time.

3.3.1 Social Sensors for Local Event Retrieval

Our motivation stems from the fact that the communities of users in Twitter often share messages about local events as they progress [66]. To give the reader a concrete example of how local events are reflected in social media, we plot in Figure 3.5 the volume of tweets that are posted within London and contain the phrase "beach boys" over a period of 12 days, where "beach boys"

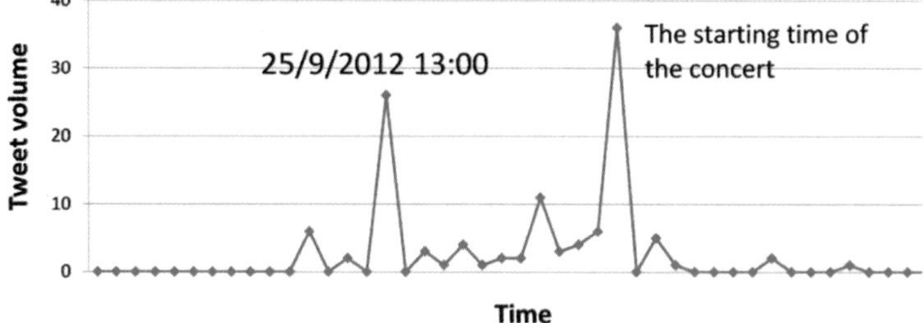

Figure 3.5 A plot of the volume of tweets in London that contain the phrase "beach boys" over time.

is the name of a rock band who held a concert in London's Royal Albert Hall during the considered time period. We observe that just before and during the concert, tweets mentioning the "beach boys" within London have spiked. This is an indication that the concert as a real-world event has been reflected in the tweeting activities within the city.

Recently, there have been some attempts to harness social media for event-based information retrieval (IR). This includes (i) identifying social media content relevant to *known* events [10, 54] and (ii) detecting *unknown* events using user-generated content in social media [11, 45, 55]. In the first case, social media content is identified to provide users with more information about a planned event (e.g. a festival or a football match). Users would be able, for example, to access tweets about ticket prices before the event, or Flickr photos posted by attendees after the event. The second case is more challenging as there is no prior knowledge about the events. While some approaches have focused on detecting news-related events [55], or simply clustering social media content based on a database of targeted events [11], a recent work has devised methods for retrieving global events from Twitter archives that correspond to an arbitrary query (event type); a problem which the authors called "structured event retrieval" over Twitter [45].

Unlike [45], which focused on non-local events, we make use of the opportunities that social media can bring to *local* search services. In particular, we define a new *localised* IR task that extends the aforementioned structured event retrieval task introduced in [45]. The task we propose aims at identifying and ranking local events based on social media activities in the area where the events occur. In other words, we use social media as a *social sensor* to detect local events in real-time.[6]

The work presented here advances the state-of-the-art in detecting and locating unknown events in social media and proposes a new IR task of local event retrieval, which is described next.

3.3.2 Problem Formulation

Our overall goal is to identify and rank local events happening in the real-world as a response to a user query. For a formal definition of a local event, we adopt a definition that has been previously used in the new event detection broadcast news task of the TDT (Topic Detection and Tracking) evaluation forum.

[6]Treating social media as a social sensor has also been suggested in previous work, for example [54] and the EU FP7 social sensors project http://www.socialsensor.eu

This definition states that an event is something that occurs in a certain place at a certain time. Formally, we consider a set of locations $\mathcal{L} = \{l_1, l_2, \ldots\}$ that are of interest to the user. The granularity of locations can vary from buildings and streets to entire cities. For example, we might consider each location to represent an area in a city in which the user is located. The city in this case is considered to be divided into equally sized areas specified by polygons of geographical coordinates, or we can use the divisions defined by the local authority such as postcodes or boroughs. Each location l_i at a certain time t_j is denoted by the tuple $\langle l_i, t_j \rangle$. We define the problem of local event retrieval as follows. For a user interested in local events within locations $\mathcal{L}$ (explicitly defined or implicitly inferred from the current user's location), the event retrieval framework aims to score tuples $\langle l_i, t_j \rangle$ according to how likely t_j represents a starting time of an event within the location l_i that matches the user query. An event is considered relevant if it matches the explicit query of the user and/or the implicit context of the user (the time of the query, the location of the user and or her profile). In other words, the event retrieval framework defines a ranking function that gives a score $R(q, \langle l_i, t_j \rangle)$ for each tuple $\langle l_i, t_j \rangle$ with regards to the user's query q. Examples of events to retrieve include festivals, football matches or security incidents. When expressed explicitly by a user, a *query* is assumed to be in the form of a bag of words (e.g. "live music", "conference").

When using Twitter as a social sensor, a location l_i at a certain time t_j is characterised by the tweeting activities observed at that location within a given timeframe $(t_j - t_{j-1})$. The tweeting activities are represented with a set of tweets originating from that location shared publicly within the given timeframe $(t_j - t_{j-1})$. This set of tweets is denoted by $T_{i,j}$. Note that the fixed timeframe is defined using an arbitrary sampling rate $\theta; \forall j : t_j - t_{j-1} = \theta$. An event happening in the real-world is represented by a tuple $\langle l, t_s, t_f \rangle$; where l is the location where the event is taking place, t_s is the starting time and t_f is the finishing time. Our aim is to use the tweeting activities as the main source of evidence to define the ranking function $R(q, \langle l_i, t_j \rangle)$. More specifically and to define the ranking function, we use the set of tweets $T_{i,j}$, and a time series of tweets $\mathcal{T}_{i,j} = \langle .., T_{i,j-2}, T_{i,j-1}, T_{i,j} \rangle$ in the location l_i before the current time t_j. This allows us to identify sudden changes in the tweeting activities, which may have been triggered by an occurrence of an event. Moreover, the event retrieval framework can identify a subset of the tweet set $T_{i,j}$ that matches the query, which may help the user in the event information seeking process.

3.3.3 A Framework for Event Retrieval

The framework aims to define an effective ranking function that scores tuples of time and location according to how likely they represent the starting time and the location of a relevant event for a given query. Note that with regards to the previous definition of the local event retrieval problem in Section 3.3.2, as a first step, we are not aiming to determine the finishing time of an event. As discussed in Section 3.3.2, here we aim to use tweets as the main source of evidence to score the tuples. In particular, we define two components built on this evidence:

1. The first component is based on the intuition that social media may reflect real-world events, hence when an event occurs somewhere we expect to find topically related social posts about it originating from the location where it occurs. To instantiate this component, for each location at a given time, i.e. for each tuple $\langle l_i, t_j \rangle$, we measure how much the tweets $T_{i,j}$ corresponding to the tuple are topically related to the query q.
2. The second component is based on the intuition that events trigger an increasing tweeting activity [66] causing peaks of tweeting rates during the event (bursts). For this component, we aim to quantify the change in the tweeting rate, the volume of tweets over time, observed at $\langle l_i, t_j \rangle$ when compared to previous observations over time at the same location. In other words, we aim to measure the unusual tweeting behaviour that may indicate an occurrence of an event. To compute the tweeting rate, we can either consider all the tweets posted within the given timeframe at the given location or only a subset of those which are relevant to the user query, e.g. tweets which contain terms of the query.

Following this, the ranking function can be defined as a linear combination of the previous two components as follows:

$$R(q, \langle l_i, t_j \rangle) \propto (1 - \lambda) \cdot S(q, T_{i,j}) + \lambda \cdot E(q, \langle l_i, t_j \rangle) \qquad (3.1)$$

where $S(q, T_{i,j})$ is the score of the tweet set $T_{i,j}$ that quantifies how much they are topically related to the query q; $E(q, \langle l_i, t_j \rangle)$ is a score proportionate to the change in the tweeting rate with regards to the query q at the given time t_j within the location l_i, and $0 \leq \lambda \leq 1$ is a parameter to control the contribution for each component in the linear combination in Equation (3.1). Next, we show how we approach the problem of quantifying each component.

Aggregating Tweets

To estimate $S(q, T_{i,j})$ in Equation (3.1), we propose to borrow ideas and techniques originally designed for the IR problem of expert search. In expert search, a profile of an expert candidate is typically represented by the documents associated to the candidate [8, 41]. Similarly, the tuple $\langle l_i, t_j \rangle$ is associated with a set of tweets. Inspired by [41], the score of each tuple (candidate) can be estimated by aggregating the retrieval scores (votes) for each tweet (document) associated to it. In [41], several voting techniques were used to aggregate the scores. We use the intuitive, yet effective, CombSUM voting technique, which estimates the final score of the tweet set representing a tuple (candidate) as follows:

$$S(q, T_{i,j}) = \sum_{t \in Rel(q) \cap T_{i,j}} (Score(q, t)) \tag{3.2}$$

where *Rel(q)* is the subset of tweets that match the query *q* and *Score(q, t)* is the individual retrieval score obtained by a traditional bag-of-words ranking function, e.g. BM25 [53]. Higher scores represent more topically related tweets for the considered tuple.

Change Point Analysis

The problem of quantifying the score $E(q, \langle l_i, t_j \rangle)$ in Equation (3.1) maps well to *change point analysis*, a previously studied problem in the statistics literature, e.g. [34, 35]. Change point analysis aims at identifying points in time series data where the statistical properties change. It has been previously applied to detect events in continuous streams of data. For example, Guralnik *et al.* developed change point detection techniques that can accurately detect events in traffic sensor data [29]. In our case, the change point analysis can be applied on the tweeting rate in a location l_i to quantify the probability that the tweeting rate at a certain time t_j represents a change point when compared retrospectively to previous points in time $t_{j-1}, t_{j-2}, . ., t_{j-k}$. We apply the Grubb's test [27] as a change point detection technique as it is computationally inexpensive and it has been successfully applied in a similar context, namely first story detection from Twitter and Wikipedia [47]. Given a location l_i and at each point of time, e.g. on minute intervals, we maintain a moving window of size *k* points, e.g. *k* minutes, over the previous observations. We apply the Grubb's test to each moving window to determine if the tweeting rate of the last point is an outlier that stands out with respect to the tweeting rates of previous observations. With Grubb's test, r_j is an outlier if $v = (r_j - \overline{x}_{j,k})/\sigma^2 > z$, where $\overline{x}_{j,k}$ is the mean tweeting rate in the window (t_{j-k}, t_j), σ is the standard deviation of the tweeting rates in the window (t_{j-k}, t_j), and z is a fixed threshold. Note that this test gives a binary decision for each point in time. We smooth this binary decision into a normalised score and use it for the second component of Equation (3.1) as follows:

$$E(q, \langle l_i, t_j \rangle) = E_c(t_j) = 1 - e^{(\frac{-\ln 2}{z} \cdot v)} \tag{3.3}$$

where $0 \leq E_c(t_j) \leq 1$ represents a score of a change point using the Grubb's test. Note that when $v = z$, the resulting score in Equation (3.3) is equal to 0.5. As previously discussed, the tweeting rate r_j can be estimated in two different ways: (i) By simply using the volume of tweets posted in the given location within the timeframe corresponding to t_j, i.e. $r_j = |T_{i,j}|$. We call this a *query independent* (QI) tweeting rate; and (ii) By using the score of the voting technique described above, i.e. $r_j = S(q, T_{i,j})$. We call this a *query dependent* (QD) tweeting rate.

It should be noted that this framework can operate in a real-time fashion on top of the SMART architecture (Section 3.2) where social feeds are incrementally indexed such that the retrieval components are able to provide the freshest results.

3.3.4 Summary

We have devised an event retrieval framework that is capable of identifying and ranking local events in a response to a user query. In [5], we have conducted an experiment on a large collection of geo-located tweets (over 1 million) collected during a period of 12 days within London. Aligned with the tweets, we have collected, through the use of crowdsourcing, local events that took place in London from local news sources. We have evaluated the effectiveness of our framework in identifying and ranking these events through its application on the geo-located tweets. Our empirical results suggest that detecting local events from Twitter using our framework is feasible but challenging. In particular, the results show that our event retrieval framework is capable of identifying and ranking "popular" events (those found by crowd-workers and reported in the web) within a city. However, when applied on more localised events, the retrieval effectiveness of the framework degrades, possibly because of their low coverage on Twitter. To deal with this caveat, in the next section, we fuse the metadata extracted from physical IoT sensors along with the social media activity to identify topics of events happening in the real-world.

3.4 Using Sensor Metadata Streams to Identify Topics of Local Events in the City

In Section 3.3, we addressed local event retrieval by using social media activity as a sensor to detect and rank events. However, social media may only cover very popular events as users may not necessarily comment on all events taking place in the city. Therefore, physical sensors that record observations about the status of the environment can provide additional evidence about the events taking place in the city. These sensors can take the form of visual sensors such as CCTV cameras, acoustic sensors such as microphones or possibly environmental sensors.

There is a wealth of research on identifying low-level human activities from acoustic and visual sensors. Often, this involves sensor signal processing to extract sensor features for modelling human activities. For example, Atrey *et al.* [7] developed a Gaussian Mixture Model using a variety of features derived from audio signal processing to classify human activities into vocal classes, such as talking and shouting, and non-vocal classes, such as walking and running. Similar approaches also used audio signal features to identify low-level human activities that are related to security incidents, such as breaking glass or explosions [25]. In addition to using acoustic sensors, several studies have been conducted to identify low-level human activities from videos. Since its introduction in 2002, the TRECVID evaluation campaign [49] has tackled a variety of content-based retrieval tasks from video recordings to support video search and navigation. This includes the semantic indexing of video segments, whereby videos are mapped to concepts, which can be certain objects or human activities [49]. Another related task is multimedia event detection, where the aim is to identify predefined classes of events in the videos. In this task, the existing effective approaches employ classifiers trained with motion features from the videos [50]. Moreover, classifying human interactions identified in video recordings has been studied to detect surveillance-related incidents [18].

Although the aforementioned approaches derive useful semantics about the multimedia content, they only consider low-level human activities. In other words, they provide sensor metadata describing low-level human activities in the physical environment. However, to the best of our knowledge, no previous work has investigated combining these sensor metadata to detect and retrieve higher level complex events taking place in the city, such as music concerts or entertainment shows, which may involve several lower level human actions. Here, we propose an approach for combining sensor metadata streams to support local event retrieval. We devise a supervised machine learning approach that combines sensor metadata to identify the topic of a potential event happening at a particular time in a certain location of the city. The topic corresponds to a set of terms representing a type of events, such as a concert or a protest. Our approach uses features from acoustic, visual and social sensor metadata. We also incorporate background features from past observations to model events that exhibit cyclic patterns such as traffic jams at peak times.

In Section 3.4.1, we define the problem of event topic identification that we tackle. This work makes use of sensor observations that are described in Section 3.4.2 – i.e. analysed video and audio recordings from two vibrant locations in the centre of a major Spanish city over a period of two weeks. Then to address the event topic identification problem, we discuss a supervised approach with two steps. In the first step, as described in Section 3.4.3, we obtain event annotations on the video and audio recordings dataset. In the second step (Section 3.4.4), we use the obtained annotations to map typical events taking place in those locations into coherent topics using a topic modelling technique.

3.4.1 Definition of Event Topic Identification Problem

The aim here is to combine the sensor metadata observations captured at different locations in a city to identify topics of potential high-level events taking place within certain locations. Formally, for a location l_i in a city, we denote the sensor metadata observations captured at time t_j in that location l_i by the vector $\overrightarrow{N}_{\langle l_i,t_j\rangle}$. The sensor metadata observations may include the crowd density identified from captured videos in the location, low-level audio events identified from the acoustic sensors installed in the location or social media activities, such as tweets posted by people at the location. The problem of event topic identification is to use the vector $\overrightarrow{N}_{\langle l_i,t_j\rangle}$ to map the tuple of time and location $\langle l_i, t_j\rangle$ to a certain topic $p_x \in \mathcal{P}$ described by a set of terms T_x; where $\mathcal{P}$ is a set of predefined topics.

In the previous section, the textual content of public tweets has been used as the only source of sensor metadata observation to identify topics of local events. Although this has worked well on popular events that attract social media activities, it does not work as well on more localised events that may not attract coverage on social media [5]. To alleviate this shortcoming, we introduce *physical* sensor metadata streams that can provide an additional evidence for the topic of an event, namely video and audio metadata observations. However, this requires understanding the semantics of visual scenes or audio recordings, which remains an open challenge. Indeed, there is no known taxonomy that maps sensor metadata to topics of high level events. To address this challenge, we propose to learn the topic associated with a tuple $\langle l_i, t_j \rangle$ from *labelled* training data using features extracted from the sensor observations $\overrightarrow{N}_{\langle l_i, t_j \rangle}$.

For this purpose, and to collect labelled training data, we obtain event annotations on a pool of videos that are identified as potential candidates to contain events. Furthermore, to extract a predefined set of coherent event topics, we apply topic modelling on the descriptions of the annotated events. We detail the event annotation and the topic extraction in Section 3.4.3. Next, we describe the sensor data collection.

3.4.2 Sensor Data Collection

Our study considers two locations in the city centre of Santander in Spain. The first location is the geographical and business heart of the city; it is a major square opposite to the municipality building. The second location is a popular open market in the city, where hundreds of people go every day for shopping, located behind the municipality building. Both locations represent vibrant and busy areas, where we expect to observe high-level events of interest such as music concerts, entertainment shows or even protests. The data collection occurred during October 2013 in both locations using an edge node deployed in Santander (see Section 3.2.1).

Table 3.1 provides a summary of the sensor data collection and the metadata produced by processing the output from the microphones and the camera in each location. For producing the audio metadata, a supervised classifier using feed forward multilayer perceptron network and low-level audio features, such as those described in [20], was developed for each of the following 6 audio classes described in Table 3.1: “crowd”, “traffic, “music”, “applause”, “speaker”, and “siren”. For video metadata, the video was processed for crowd analysis where we calculate the crowd density, in desired areas, by estimating the foreground components of the video. In addition to the acoustic and visual sensors, we collected parallel social media activity in the city. In particular, using the Twitter Public Streaming API[7], we obtained tweets related to each location (as identified by their geo-locations).

[7]https://dev.twitter.com/streaming/public

Table 3.1 Summary of sensor data collection

Locations	2 (square & market)
Physical sensors	A camera and microphones (in each location)
Raw output	1600 × 1200 video @ 20 fps 16 Khz audio @ 64 kbits/s (audio is multiplexed with the video)
Audio metadata	classification scores for 6 audio classes (i) "crowd": noise from a crowd of people (ii) "traffic": car and road noises (iii) "music": music played outdoors (iv) "applause": applause, yelling or cheering (v) "speaker": speech over loud speakers (vi) "siren": noise of police cars & ambulances
Video metadata	crowd density in the scene
Twitter	geo-tagged tweets within each location

3.4.3 Event Pooling and Annotation

In this section, we describe our approach for obtaining event annotations on the recordings collected from the two used locations. Recall that our ranking units (the documents) are tuples of time and location. Each tuple represents a segment of recordings at a location. The length of the segment, the sampling rate to obtain the tuples, can be predfined and we follow [5] in setting the sampling rate to 15 minutes. Coarser- or finer-grained sampling rate can be investigated in future work for different types of events e.g. emergency events may require a finer-grained sampling.

For annotation, we consider a period of 2 weeks starting from 19 October 2013, around a week after the start of the data collection (11 October 2013) to allow the estimation of background features. Since it is expensive to examine all recordings and annotate them with events, we employ a pooling approach [16], as commonly used in IR (Information Retrieval) evaluation forums, such as TREC. For pooling, we identify candidate segments of videos where high-level events may have occurred by applying the change component of the event retrieval framework described previously (c.f. Section 3.3). In particular, the change component of this framework identifies segments where sensor metadata observations change unusually in a location, e.g. unusual change in crowd density. We use 4 different types of sensor metadata observations to generate the pool (a subset of those listed in Table 3.1): (i) the median values of the video crowd density, (ii) the median values of the crowd audio classification score, (iii) the median values of the music audio classification score, and (iv) the total number of tweets posted. As a result, we obtain a total of 155 candidate segments. The video recording software produced videos with lengths of either 30 minutes or 1 hour, and the total number of video recordings that correspond to the 155 segments are 69 videos.

The generated candidate segments of videos were then annotated by two groups of human annotators, English and Spanish annotators, who were asked to examine the videos, describe events that they observe by typing in terms, and rate their intensity on a 3-point scale (Low, Medium, and High) according to how likely they are to generate public interest. The intensity is akin to graded relevance used in traditional IR evaluation approaches [59]. We provided the annotators with a web-based interface, of which we show a snapshot in Figure 3.6.

Statistics of the obtained annotations are summarised in Table 3.2. From the last row it can be observed that we obtain a total of 55 annotated videos, of which 21 were annotated by more than one annotator. The agreement between annotators is estimated by converting the intensity levels to binary decisions,

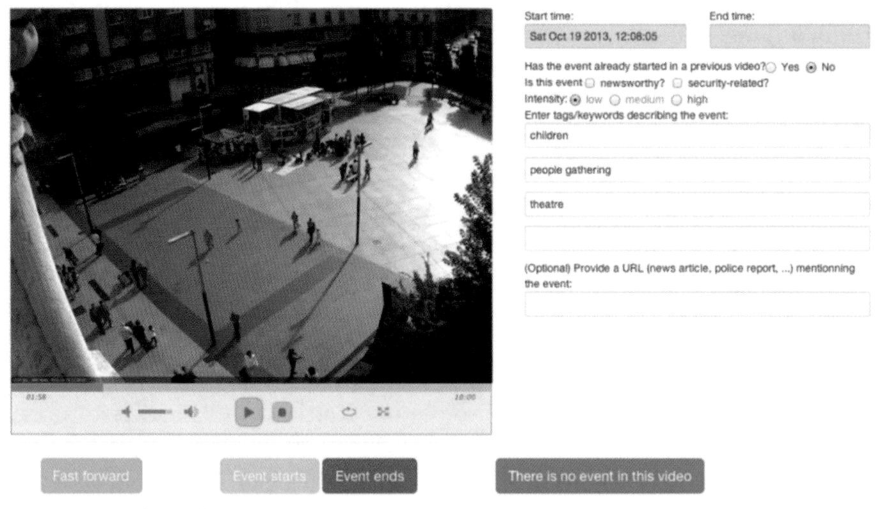

Figure 3.6 Components a snapshot from the annotation interface.

Table 3.2 Statistics from annotated videos

Annotators	Unique Videos	Annot. Ratio	Mutliple Annotations	Agreement
4 English	29	29/69 = 42%	1 video	100%
5 Spanish	47	47/69 = 68%	13 videos	77%
Both	55	55/69 = 79%	21 videos	71%

using "Medium" as a threshold. We observe that a reasonable agreement is achieved in all cases (lowest is 71%), which gives us confidence in the annotations obtained.

For the set of annotated pooled segments, we obtain terms describing events that were identified in these segments. For each annotated segment, we construct a virtual document that consists of all of the terms provided by the annotators. Since the pooled videos were annotated by both Spanish and English annotators, these virtual documents are bilingual and contain English and/or Spanish terms. Figure 3.7 presents word clouds for frequent terms occurring in the English (a) and Spanish (b) annotations, respectively, where word size is indicative of frequency in the annotations.

To cluster events into various topics, we propose to use topic modelling on the document collection of all constructed virtual documents of terms. We use the Latent Dirichlet Allocation (LDA) topic modelling implemented in the Mallet toolkit [43]. In Table 3.3, we list the top terms of 7 identified topics from the English annotations only. From the table, we can observe that the identified topics are reasonable where we see some interesting associations of terms that describe typical high-level events taking place in the square and the market, e.g. 'demonstration' and 'show' in topic 4, and 'children' and 'entertainment' in topic 6.

3.4.4 Learning Event Topics

In this section, we discuss our supervised approach for event topic identification, where the aim is to identify the topic of a segment $\langle l_i, t_j \rangle$ using the sensor metadata observations $\overrightarrow{N}_{\langle l_i, t_j \rangle}$. To train our supervised approach, we construct a *labelled* dataset of event topics from the annotated video pool collected. The labelled data consists of segments (tuples of time and location) labelled with

(a) English. (b) Spanish.

Figure 3.7 Word clouds for frequent terms occurring in the English (a) and Spanish (b) annotations, respectively, where word size is indicative of frequency in the annotations.

Table 3.3 Topics identifed with topic modelling using the english annotations

Topic	Top Terms of the Topic
Topic 1	loudspeaker people fanfare police drums procession
Topic 2	microphone rings speech public claps
Topic 3	gathering plaza people booth theatre music
Topic 4	demonstration sitting event sound speak show
Topic 5	market protest cars ongoing children fair
Topic 6	children people shopping middle entertainment
Topic 7	music singing playing guy bells whistles

either an event topic or with the label 'no event' indicating that no event of interest has occurred in the corresponding time and location. We labelled each annotated segment in the pool to the most probable topic according to the LDA topic modelling configured by setting the number of topics to 7.[8] Unlabelled segments or where the annotators did not identify any event are associated to the 'no event' label. To illustrate the volume of the data and the distribution of labels, we detail in Table 3.4 the number of segments for each label when using topic modelling on all Spanish and English annotations and setting the number of topics to 7.

We consider the problem of identifying the topic of a pooled segment as a classification task. Using the constructed labelled data, we train a binary classifier for each of the labels with features derived from various sensor metadata streams. Our intuition is that such labelled data would allow us to learn the semantics of a combination of sensor metadata. In other words we aim to match sensor metadata to topics defined using the annotations. For training the classifier, we investigate two main sets of features for the segment, *observation features* and *background features*. Table 3.5 summarises those features. The *observation features* are extracted from the sensor metadata observed in the location and time corresponding to the segment. The *background features* aim to model past observations and cyclic patterns of activities that take place over time in the same location. The intuition is that some events

Table 3.4 Distribution of labels

Lab.	#	Lab.	#	Lab.	#	Lab.	#
top.1	12	top.3	2	top.5	0	top.7	11
top.2	8	top.4	32	top.6	24	no event	66

[8]We use 7 topics since we have observed that with this setting we obtain the most coherent topics after experimenting with other alternatives (varying the number of topics between 5 and 10).

Table 3.5 Features devised for topics identification

8 Observation Features		
Audio features	6	median of the classification score for each audio class (crowd, traffic, music, applause, speaker, siren)
Video features	1	median of the crowd density score
Twitter features	1	number of tweets geotagged within the location in the past one hour
16 Background Features		
Daily aggregates	8	for each of our 8 observation features its daily median from all available past observations at the same time from previous days
Weekly aggregates	8	for each of our 8 observation features its median from all available past observations at the same time on the same day of previous weeks
Total	24	

are periodic and exhibit a long-term pattern, e.g. traffic jams at peak times resulting in a high traffic audio classification score, or entertainment shows taking place in the square at the same time on the weekends. Modelling cyclic patterns, i.e. daily and weekly cycles, from the sensor metadata observations would enable the supervised classifier to identify recurring background events or noise which are not of interest such as traffic jams. Similarly, it would help to identify recurring events of interest such as entertainment shows.

Using the labelled dataset of segments along with the features described in Table 3.5, we apply supervised machine learning to learn a binary classifier for each label. In particular, we experiment with Random Forests [15] as a learning algorithm.[9] Next we conduct a number of experiments to evaluate the accuracy of our classifier and the effectiveness of the various devised features.

3.4.5 Experiments

To evaluate our approach for identifying the topic of a candidate segment, we use the dataset of labelled segments described in Section 3.4.4. We perform a 10-fold cross validation and report the average accuracy across all labels (a label for each topic and the label 'no event'). In addition to using different instantiations of our classifier, we also compare our classifier to an alternative baseline. The "majority" baseline assigns the most common label in the

[9]We also experimented with other supervised machine learning algorithms, such as naive Bayes and SVM, however we only report results with Random Forests since they achieve the best performances and the conclusions with other algorithms are similar.

training data to the segments in the testing data. Table 3.6 summarises the results.

We observe from the table that all instantiations of our approach are markedly better than the majority baseline. In particular, when using only the observation features our approach achieves an F_1 accuracy of 0.686. We also observe that this performance further increases when using the background features. Indeed the best performance is achieved when using all background features along with the observation features ($F_1 = 0.766$). This illustrates that modelling cyclic patterns by aggregating sensor metadata from previous observations helps in better identifying whether a candidate segment represents an event and in identifying the topic of an event.

Furthermore, we conduct an ablation study to identify which features are more effective for topic identification. We remove one of our 8 observation features when learning the classifier. We report the results in Table 3.7. For example, the row headed "– (Audio crowd)" means that we use all the observation features apart from the audio crowd score. We observe that removing any of the features results in a degradation of performance for accuracy and precision. This is an interesting observation and highlights the importance of having rich metadata describing the environment for identifying the topics of high-level events. However, we also observe that the performance

Table 3.6 Performance of topic identification

Approach	F_1 Accuracy	Precision	Recall
Majority baseline	0.254	0.181	0.426
Obs. Feat.	0.686	0.705	0.697
Obs. & Daily	0.740	0.759	0.761
Obs. & Weekly	0.715	0.715	0.729
Obs. & All background	**0.766**	**0.781**	**0.762**

Table 3.7 Results of the ablation study

Model	F_1 Accuracy	Precision	Recall
All observation features	**0.686**	**0.705**	**0.697**
– (Audio crowd)	0.635	0.624	0.635
– (Audio traffic)	0.681	0.678	0.691
– (Audio applause)	0.680	0.678	**0.697**
– (Audio music)	0.685	0.682	**0.697**
– (Audio speaker)	0.657	0.656	0.665
– (Audio siren)	0.656	0.655	0.665
– (Video crowd)	0.652	0.651	0.665
– Twitter	0.682	0.677	0.697

degrades most when removing the audio crowd score and the crowd density features. This suggests that the crowd level, as detected by the acoustic or visual sensors, is important to identify events and to distinguish their topic.

3.4.6 Summary

In summary, we described an approach for fusing sensor metadata streams to identify the topics of events, happening within a city, building on the SMART framework (described previously in Section 3.2). In particular, this approach trains a classifier to identify event topics from candidate segments of audio and video recordings. Experimental results demonstrate that the best accuracy for event topic identification can be achieved by combining features from a variety of diverse sensors (acoustic, visual and social). This shows the advantages "that the social and other IoT stream fusion brings to event topic identification. Moreover, it paves the way towards more effective local event retrieval that harness both physical and social sensor streams in cities with significant IoT-connected sensing infrastructures, by combining the visions from Section 3.2 (an infrastructure for searching IoT), Section 3.3 (location event retrieval from social streams) and event topic identification from physical and social sensor streams.

3.5 Venue Recommendation

The advances of smartphone devices and wireless communication technologies have enabled people to search for information in almost every situation, and no longer simply when at a desk. However, as the information on the internet has dramatically grown every day, searching for relevant information seems to be a difficult and time-consuming task for instance, due to the cognitive complexities of expressing information needs by typing on a smartphone screen. For this reason, recommendation systems have become ubiquitous tools to obtain information, by predicting what the user wants without the need for an explicit query.

In recent years, Location-based Social Networks (LBSNs) have emerged, such as Foursquare[10] and Yelp[11], which enable users to search for Points-of-Interest (POI) or venues[12], share their physical location (check-ins[13]) as well

[10]https://foursquare.com/

[11]http://www.yelp.co.uk

[12]We use the term venue or POI interchangably.

[13]A term used by Foursquare to denote users sharing their current location with the LBSN.

as rate and comment after visiting a POI. Moreover, other users may consider those ratings and comments to select the POIs to visit at a later time. The recommendation of appropriate POIs to users, e.g. a restaurant they are likely to visit, has become an important feature for LBSNs, which assists people to explore new places and helps business owners to be discovered by potential customers.

Venue recommendation is an example of a recommendation task: given no explicit 'query' by the user, but knowledge about the user's preferences and about the venues, can a system predict which venues the user may wish to visit? The types of information that are available for this task are summarised in Table 3.8. For instance, if the user has checked-in or rated other venues before, then this provides user preference information, which can be used by collaborative filtering approaches to suggest venues of interest (discussed in details in Section 3.5.1 below).

Nevertheless, information about each venue itself can help to predict its likely suitability. For instance, the Foursquare website lists a city park as a top nearby venue, regardless of the time of day, when (say) late in the evening that park may be both closed and unsuitable to visit. Therefore IoT and social sensing technology have a role to play in predicting the occupancy of venues. Predicted occupancy and similar measures of popularity are examples of venue-dependent features (i.e. which are the same for all users) – and are discussed further in Section 3.5.2.

Finally, the *contextual* situation of the users when requesting venue recommendations can also have an impact on the appropriate choice of venues: clearly, context can encapsulate the location of the user – as nearby venues are more likely to be useful to the user; however, the people they are with (alone, with colleagues, family or friends) may also significantly impact upon the most appropriate choice of venues. In Section 3.5.3 we highlight recent work in context-aware venue recommendation.

In the remainder of this section, we discuss recent research in venue recommendation, particularly highlighting our own work, which builds upon

Table 3.8 Sources of data for venue recommendation

	Example Sensors	
Information Type	Physical	Social
Venue	# cell phones nearby	# recent check-ins
	# subway exists nearby	# comments
User Preference	User's distance to the venue	The user likes similar venues
		The user has commented positively about a similar venue

the SMART architecture (described in Section 3.2). The three different sources of evidence, and how they are modelled, are described in turn: user preferences (e.g. through collaborative filtering approaches) (Section 3.5.1), venue dependent evidence (Section 3.5.2), and contextual preferences (Section 3.5.3).

3.5.1 Modelling User Preferences

In terms of modelling the preferences of users, collaborative filtering (CF) is a widely used technique to generate personalised recommendations. CF typically exploits a matrix of user-venue preferences in order to generate venue recommendations for individual users. There are two major categories of traditional CF approaches namely memory-based CF and model-based CF [1, 26]. The memory-based CF approaches are categorised as user-based or venue-based. A typical user-based CF approach predicts a user's rating on a target venue by aggregating the ratings of K similar users who have previously rated the target venue. The similarity between two users is usually identified using the Pearson correlation or the cosine similarity upon their rating vectors [57]. Intuitively, the user-based CF approaches assume that users who share similar preferences will like the same venue e.g. I like what my friends like. The extension of user-based CF approaches has been shown to improve the quality of recommended venues such as through the introduction of fine-grained neighbour-weighting factors [32] or by exploiting a recursive neighbour-seeking scheme [65]. In contrast, venue-based CF approaches suggest venues on the basis of information about other venues that a user has previously rated [21]. The suggestion of venues for a given user are ranked by aggregating the similarities between each candidate venue and the venues that the user has rated. Although typical memory-based CF approaches have been shown to be effective in suggesting venues to users, the main drawback of such approaches is that the computation of similarities between all pairs of users or venues is expensive due to its quadratic time complexity. Moreover, as memory-based CF approaches are dependent on the availability of human ratings, the effectiveness of these approaches significantly decreases when they are faced with sparse ratings.

On the other hand, model-based CF approaches were introduced to address the shortcomings of memory-based CF approaches [14, 9]. Such approaches are based on supervised models which are trained on the user-venue matrix [1, 26]. The trained prediction models can then be used to generate suggestions for individual users. Recently, the most well-known technique of model-based CF approaches is matrix factorization (MF) [36]. The advantages of

MF techniques are their scalability and accuracy. Generally, MF models learn latent features of users and venues from the information in the user-venue matrix, which are further used to predict new ratings between users and venues.

Various recent works on venue recommendation have exploited user-generated information (e.g. check-in and venue information) from LBSNs. Such approaches typically apply widely-used CF techniques to suggest personalised venues to users. For example, friend-based CF approaches can recommends POIs to visit based on collaborative ratings of venues made by the user's friends [40, 62]. Yang *et al.* [61] proposed a model that estimates a venue's quality based on a sentiment score calculated based on the tips (comments) made by users in the LBSN, and then recommended venues based on this sentiment score.

Even if there is no information available about which venues a user has previously visited, other proxy information can still be obtained with which to personalise the suggestion of venues. For instance, one venue recommendation approach that we have proposed in [23] used the users' Facebook profiles to permit personalisation, even when the venues being suggested were from the separate Foursquare LBSN. In general, we use a probabilistic model to describe the preferences of a user determined from Facebook – as well as the likely interests of users – in terms of a coarse-grained ontology from the Open Directory Project (DMOZ.org): e.g. Arts, Games, Health, Technology etc. In particular, we examine the entities liked by users on Facebook to build up a preference distribution over the DMOZ categories. However, as the Foursquare entities may only have a single name to describe them, this may be insufficient information to accurately predict which DMOZ categor(ies) these entities should belong to. To alleviate this problem, we issue each Facebook entity's name as a query to a web search engine, and analyse the contents of the returned pages to determine which categories they belong to. This allows a model of the users' preferences to be determined based on their Facebook' Likes.

Similarly, when considering a venue, we determine which categories that it should belong to by also issuing the name of the venue to a web search engine. Figures 3.8 (a) and (b) pictorially depict the aforementioned processes for the Facebook' Likes and the venues, respectively. Then, the personalisation of venues suggested to a user can be achieved by suggesting venues with more similar category distributions to that of the users. Through a user study involving 100 users and three different cities (Amsterdam, London, San Francisco), we evaluated our complete probabilistic model [23]. Our findings suggested that while our personalised model was effective, it was residents

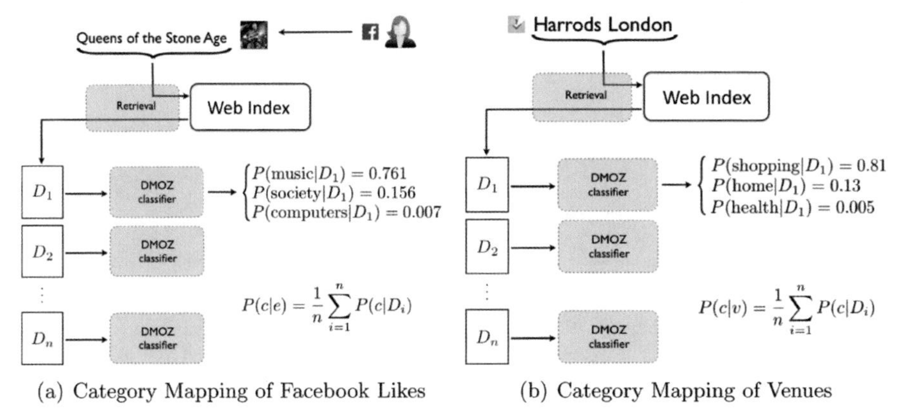

(a) Category Mapping of Facebook Likes (b) Category Mapping of Venues

Figure 3.8 Obtaining DMOZ category distributions from Facebook' likes and venues.

rather than visitors to a city who were found to most prefer personalised venue suggestions.

Overall, we have shown how personalised venue suggestions can be achieved, but there are other factors that may cause a venue to be selected by a user or not – indeed, our own user study in [23] found that tourists were more attracted by popular venues, particularly during the evening. In the next section, we describe how measures of the venue's popularity can be sensed, both physically or socially.

3.5.2 Venue-dependent Evidence

The act of checking-in on a LBSN such as Foursquare provides a number of signals about a venue – the short-term popularity of a venue, as well as an aggregate signal about its popularity at this time, as well as at the current day of the week and season of the year. IoT-connected sensors that can detect a busy venue, such as through CCTV analysis and/or audio analysis (see Table 3.8) can also similarly be used.

To predict the attendance of a venue, we constructed a time series of attendance for each individual venue [23]. Time series are numerical information that are observed sequentially over time. By obtaining the number of people currently visiting the venue from Foursquare every hour for each venue, we are able to build a comprehensive time series of venue attendance. Figure 3.9 shows how a state-of-the-art neural network-based approach can predict the attendance of the famous Harrods department store in Knightsbridge, London. Using such predicted occupancy figures for venues improves the effectiveness of a venue suggestion approach [23] – as illustrated in Table 3.9.

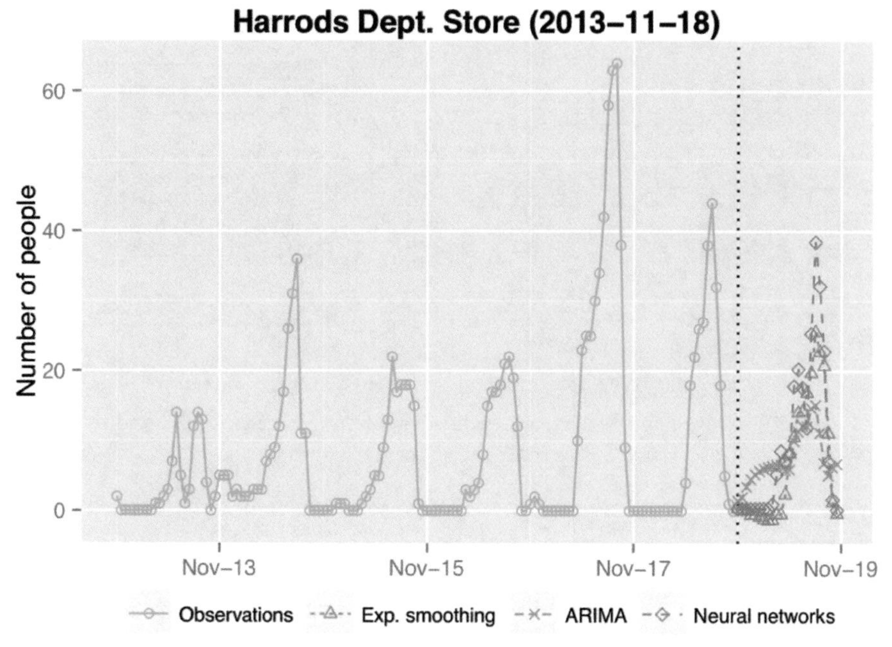

Figure 3.9 Predicting occupancy from Foursquare check-in time-series.

Table 3.9 Examples of venue recommendations produced by our model for user in a central location in London at two different times

Friday 03 April 14:00	Sunday 5 April 00:00
Debenhams	Novikov Restaurant & Bar
Natural History Museum	Boujis (nightclub)
Selfridges & Co	
National Gallery	
Apple Store	
London Victoria Railway Station	
Victoria and Albert Museum (V&A)	
Millbank Tower	
Science Museum	
Piccadilly Circus	

Of course, there are other sources of evidence in a LBSN that are indicative of a venue's popularity and hence a priori its suitability for recommendation to any user. In a separate work, we examined a number of venue-dependent features for making effective venue suggestions, such as the number of check-ins, number of photos, average ratings etc. [22] (summarised in Table 3.10).

To create an effective venue suggestion model, these features were combined with user-venue features (which model the user's venue preferences, in order to make personalised suggestions) within a learning-to-rank approach. In doing so, the application of the learning-to-rank approach [22] aims to find a combination of the features that can best satisfy users, determined by a set of training observations (users with known venue preferences). The experiment made use of the state-of-the-art LambdaMART learning-to-rank approach [60], which is an adaptation of gradient boosted regression trees to make effective rankings.

Figure 3.10 shows how performance can increase or decrease when venue-dependent features are removed from the model. This takes the form of an *ablation* experiment to explore the individual effectiveness of these venue-dependent features, in order to determine which single features are the most effective when suggesting venues to users. In this experiment, we consider the LambdaMART ranking model – learned using all features – as a baseline, and we compare its performances to other LambdaMART models that have been learned after removing each of the venue-dependent features individually – a decrease in performance implies that the feature is deemed useful.

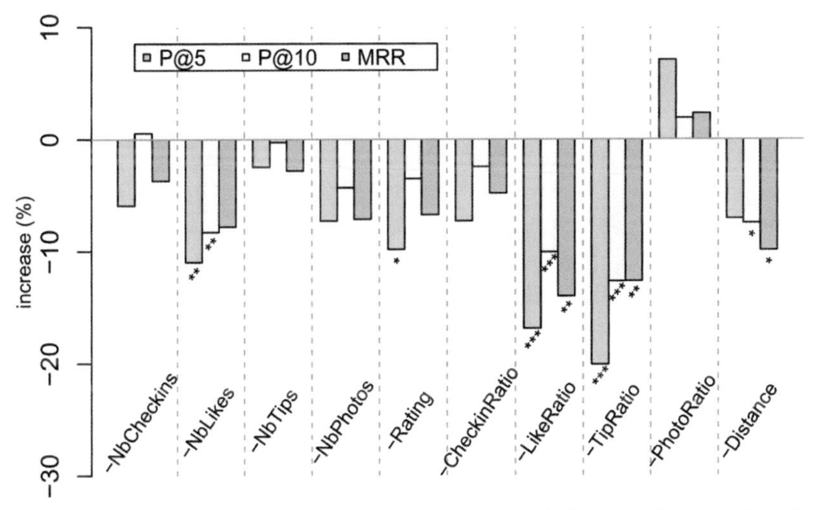

Figure 3.10 Percentage of improvement obtained when independently removing single venue-dependent features, with respect to a LambdaMART baseline that uses a total of 64 features. Improvements are expressed in terms of P@5, P@10, and MRR. Statistical significance is stated according to a paired t-test (*: $p < 0.05$, **: $p < 0.01$, ***: $p < 0.001$).

Source: [22].

Table 3.10 Venue-dependent Foursquare features used by Deveaud et al. [22]

Feature Name	Description
NbCheckins	Total number of check-ins in the venue.
NbLikes	Total number of "likes" for the venue.
NbTips	Total number of "tips" (comments) for the venue.
NbPhotos	Total number of photos for the venue.
Rating	Average of all the ratings given by the users for the venue.
CheckinRatio	$\frac{\text{NbCheckins}}{\text{NbCheckinsInCity}}$
LikeRatio	$\frac{\text{NbLikes}}{\text{NbLikesInCity}}$
TipRatio	$\frac{\text{NbTips}}{\text{NbTipsInCity}}$
PhotoRatio	$\frac{\text{NbPhotos}}{\text{NbPhotosInCity}}$
Distance	Distance of the venue from the center of the city.

On analysis of Figure 3.10, the first observation we make is that PhotoRatio appears to be harmful. When Foursquare venues do not have any photo, the value of this feature is equal to zero, which seems to confuse the learner.

Likes and tips, which are more abundant and hence do not suffer from this problem, appear to be very strong indicators of relevance. It is important to note that the raw numbers (i.e. NbLikes and NbTips) are not enough, and that using the city context greatly improves the importance of these features (see LikeRatio and TipRatio). The rating of the venue (which is an average of all the ratings provided by Foursquare users) is also a good indicator of relevance, but to a lesser extent than LikeRatio and TipRatio. Finally, the distance between the venue and the center of the city also seem to play an important role. Since city centres usually are the most vibrant parts, using this distance as a feature allows the learned model to implicitly separate potentially relevant and attractive venues from unpopular ones.

Overall, our experiment in [22] – and highlighted above – shows the importance of venue-dependent features for effective venue suggestions.

3.5.3 Context-Aware Venue Recommendations

In addition to making recommendations based on user preferences and the popularity of the venue, another area that has emerged recently is context-aware venue recommendation (CAVR, also known as contextual suggestions). CAVR acknowledges that the appropriate venues to be recommended to a user may depend on the contextual environment of the user. Context as a notion is wide-reaching, but for venue recommendation, it can encompass factors detectable about the user, such as the location of the user, the time of day, the weather, as well as human factors, such as who the user is with (colleagues, friends, partner family, etc), that they may explicitly provide to the venue recommendation system.

Various existing works have shown that considering such context and leveraging the use of user-generated data in LBSNs can significantly enhance the effectiveness of CAVR applications [37, 38, 63]. Yuan *et al.* [63] developed a collaborative *time-aware venue recommendation* approach that suggests venues to users at a specific time of the day. In particular, they leveraged the use of historical check-ins of users in LBSNs to model the temporal behaviour of the users and extend the user-based CF technique to incorporate both temporal and geographical effects using linear combination. Recently, Li *et al.* [37], proposed factorization methods for making venue recommendations, which can exploit different types of context information (e.g. the user's location and the time of the day). Previous works on CAVR (e.g. [37, 63]) used check-in data from LBSNs to evaluate the accuracy of their recommendations, by assuming that users implicitly like the venues they visited.

Since 2012, the US National Institute of Standards and Technology (NIST) have been developing reliable and reusable test collections and evaluation methodologies to measure the effectiveness of CAVR systems through the Contextual Suggestion track [19] of the Text REtrieval Conference (TREC) evaluation campaign. In particular, the task addressed by the TREC Contextual Suggestion track is as follows: given the user's preferences (ratings of venues) and context (e.g. user's location, city), produce a ranked list of venue suggestions for each user-context pair. A description of the contexts addressed in the 2015 TREC Contextual Suggestions track are presented in Table 3.11.

Table 3.11 The 12 dimensions of the contextual aspects proposed by the TREC 2015 contextual suggestion track

Aspect	Dimension	Description
	Day Time	Is a venue suitable to visit between 6:00 AM – 6:00 PM?
Duration	Night Time	Is a venue suitable to visit between 6:00 PM – 6:00 AM?
	Weekend	Is a venue suitable to visit on weekend?
	Spring	Is a venue suitable to visit between March and May?
Season	Summer	Is a venue suitable to visit between June and August?
	Autumn	Is a venue suitable to visit between September and November?
	Winter	Is a venue suitable to visit between December and February?
	Alone	Is a venue suitable to visit alone?
Group	Friends	Is a venue suitable to visit with friends?
	Family	Is a venue suitable to visit with family?
Type	Business	Is a venue suitable to visit for a business trip?
	Holiday	Is a venue suitable to visit for a holiday trip?

The availability and popularity of the TREC Contextual Suggestion track test collections has accelerated research into this challenging task. A few research groups participating in the TREC Contextual Suggestion track have attempted to explicitly model the contextual appropriateness of the venues. Hashemi *et al.* [31] applied parsimonious language models [33] to rank suggestion candidates based on the given contextual information such as trip duration and type, and information from the user's profile, such as their age and gender. Textual language models of each contextual aspect were built offline, and then the relevance of a given venue to the various contextual aspects of the user were estimated by calculating the KL-divergence of the standard language models of suggestion candidates and the language model of different contexts that was built in advance. The contextual relevance of different contextual aspects to the given suggestion candidate is trained using a pairwise SVM rank learning-to-rank model. In [44], we proposed two approaches for CAVR. First, a Factorization Machines-based approach proposed by [52] to rank the candidate venue suggestions. The factorisation machines receive as input instances that enclose the information related to a user, a venue he/she visited and the context of the visit in the form of numerical vectors. In particular, we trained the factorisation machines to reduce the error in the ranking of the user profiles by adapting the list-wise error function of ListRank [56] for their factorisation machine model. The second approach is a learning-to-rank based approach where contextual features are extracted from the user-generated data from LBSN (e.g. timestamp of comments and photo).

Recently, we proposed a supervised approach that predicts the appropriateness of venues to particular contextual aspects, by leveraging user-generated data in LBSNs such as Foursquare [42]. This approach learns a binary classifier for each dimension of three considered contextual aspects proposed by the TREC Contextual Suggestion track (see Table 3.11). A set of discriminative features are extracted from the comments, photos and website of venues. For instance, when travelling with children, the website of an appropriate restaurant may mention a children's menu; similarly, users may reminisce about pleasant times they had with their family using the LBSN comment functionality. By analysing these sources of evidence, we showed in Section 4.2 that both the websites of venues and comments left by users on the LBSN could accurately predict if a venue was suitable for the various contextual aspects.

3.5.4 Summary

Venue recommendation is an important task, for instance exploring a new city, as evidenced by the popularity of LBSNs and other websites such as

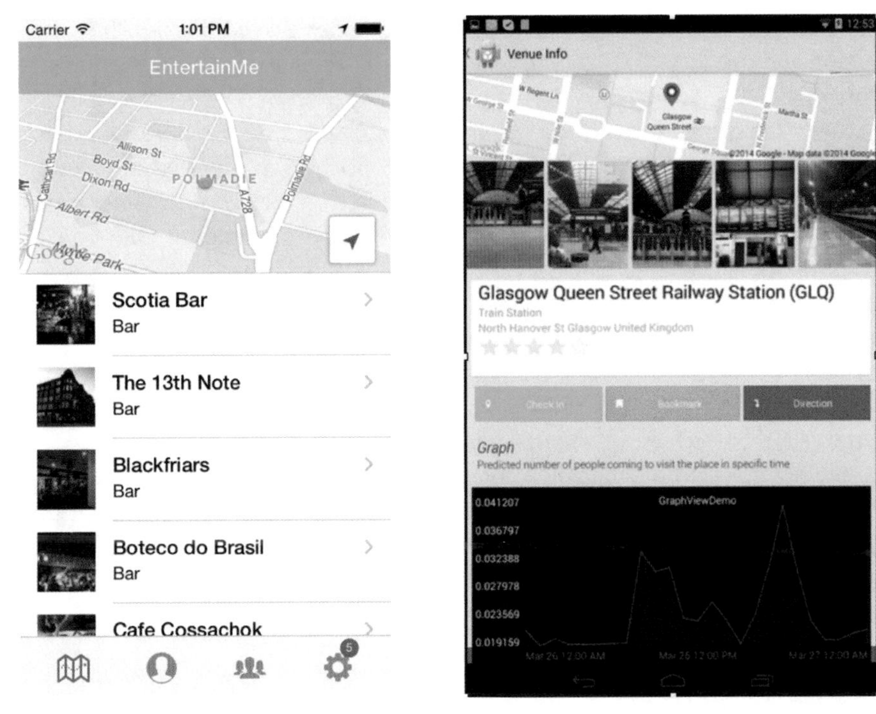

Figure 3.11 Screenshots of the EntertainMe! mobile venue recommendation application.

TripAdvisor. There is a significant body of research on venue recommendation, with new models making increasing use of social sensing. As highlighted in Table 3.8, physical sensors may assist in making recommendations, for instance, by recognising that some venues are not appropriate for poor weather conditions.

In [24], we developed a mobile application called EntertainMe!, based on the SMART architecture and employing some of the techniques described in Sections 3.5.1, 3.5.2 and 3.5.3. A screenshot of the mobile application is shown in Figure 3.11.

3.6 Conclusions

In this chapter, we described the SMART architecture, allowing to develop real-time search applications on the so-called Internet of Things (IoT) infrastructure. We illustrated the use of the SMART architecture through three applications, addressing local search, event topic identification and venue recommendation, respectively. In general, these applications show how real-time information needs by users can be better served through the integration and fusion of IoT sensing streams with social content within a unified platform.

Over the next few years, as small IoT-enabled devices become ever more ubiquitous, search platforms like SMART will become increasingly more important as a means to convert the data collected by these sensors into useful actionable information. Indeed, we expect that the continuing adoption of IoT sensors will enable a wide variety of new user information needs to be satisfied, both in the short and long term. For instance, large shopping malls are installing IoT sensing infrastructures, creating 'SMART buildings', with the aim of enhancing the user's shopping experience [51]. This allows users' positions to be tracked as they move around the mall, which could enable better personalised search results, e.g. if a user enters a query such as 'sports shirts' into their mobile then the shopping results could be augmented with the location of those products nearby in the mall. Furthermore, in the smart utility space, devices such as smart fridges can make use of platforms like SMART to suggest contextual queries/reminders to the user. For example, if the fridge detects that the user is about to run out of milk, the fridge could push search results for milk to the user on their smart phone.

The SMART (http://www.smartfp7.eu/) platform is an open source project, intended to facilitate harnessing the power of the Internet of Things infrastructure in search applications. Source code and example applications can be downloaded from https://github.com/SmartSearch.

Acknowledgements

The authors acknowledge the support of the EC co-funded project SMART (FP7-287583).

References

[1] G. Adomavicius and A. Tuzhilin. Toward the next generation of recommender systems: A survey of the state-of-the-art and possible extensions. *Knowledge and Data Engineering, IEEE Transactions on*, 17(6): 734–749, 2005.

[2] A. Agarwal, B. Xie, I. Vovsha, O. Rambow, and R. Passonneau. Sentiment analysis of twitter data. In *Proceedings of LSM '11*, pages 30–38, Portland, Oregon, USA, 2011.

[3] D. Albakour, C. Macdonald, and I. Ounis. Query scoring and anticipation subsystem. Technical Report D5.3b, SMART-FP7.eu, 2015.

[4] M. Albakour, C. Macdonald, I. Ounis, A. Pnevmatikakis, and J. Soldatos. Smart: An open source framework for searching the physical world. In *Proceedings of the SIGIR'12 workshop on Open Source Information Retrieval*, pages 48–51, 2012.

[5] M.-D. Albakour, C. Macdonald, and I. Ounis. Identifying local events by using microblogs as social sensors. In *Proc. of OAIR'13*.

[6] J. Allan, J. Carbonell, G. Doddington, J. Yamron, and Y. Yang. Topic detection and tracking. pilot study final report. In Proceedings of the DARPA Broadcast News Transcription and Understanding Workshop, 1998, pp. 194–218.

[7] P. K. Atrey, M. Maddage, and M. S. Kankanhalli. Audio based event detection for multimedia surveillance. In *Proc. of ICASSP'06*.

[8] K. Balog, L. Azzopardi, and M. de Rijke. A language modelling framework for expert *Information Processing & Management*, 45(1):1–19, 2009.

[9] C. Basu, H. Hirsh, W. Cohen, et al. Recommendation as classification: Using social and content-based information in recommendation. In *AAAI/IAAI*, pages 714–720, 1998.

[10] H. Becker, D. Iter, M. Naaman, and L. Gravano. Identifying content for planned events across social media sites. In *Proceedings of WSDM '12*, pages 533–542, New York, NY, USA, 2012. ACM.

[11] H. Becker, M. Naaman, and L. Gravano. Learning similarity metrics for event identification in social media. In *Proceedings of WSDM'10*, pages 291–300, 2010.

[12] E. Borden. Pachube internet of things "bill of rights". http://blog.cosm.com/2011/03/pachube-internet-of-things-bill-of.html, 2011.

[13] M. Botts, G. Percivall, C. Reed, and J. Davidson. Ogc sensor web enablement: Overview and high level architecture. In *GeoSensor Networks*, pages 175–190, 2008.

[14] J. S. Breese, D. Heckerman, and C. Kadie. Empirical analysis of predictive algorithms for collaborative filtering. In *Proceedings of the Fourteenth conference on Uncertainty in artificial intelligence*, pages 43–52. Morgan Kaufmann Publishers Inc., 1998.

[15] L. Breiman. Random forests. *Machine learning*, 45(1):5–32, 2001.

[16] C. Buckley and E. M. Voorhees. Retrieval evaluation with incomplete information. In *Proc. of SIGIR'04*.

[17] J. Camp, J. Robinson, C. Steger, and E. Knightly. Measurement driven deployment of a two-tier urban mesh access network. In *Proceedings of MobiSys '06*, pages 96–109, 2006.

[18] F. Chen and W. Wang. Activity recognition through multi-scale dynamic bayesian network. In *Proc. of VSMM'10.*

[19] A. Dean-Hall, C. L. Clarke, J. Kamps, P. Thomas, N. Simone, and E. Voorhees. Overview of the trec 2013 contextual suggestion track. Technical report, DTIC Document, 2013.

[20] J. Dennis, Q. Yu, H. Tang, H. D. Tran, and H. Li. Temporal coding of local spectrogram features for robust sound recognition. In *Proc. of ICASSP'13.*

[21] M. Deshpande and G. Karypis. Item-based top-n recommendation algorithms. *ACM Transactions on Information Systems (TOIS)*, 22(1): 143–177, 2004.

[22] R. Deveaud, M. Albakour, C. Macdonald, I. Ounis, et al. On the importance of venue-dependent features for learning to rank contextual suggestions. In *Proceedings of the 23rd ACM International Conference on Conference on Information and Knowledge Management*, pages 1827–1830. ACM, 2014.

[23] R. Deveaud, M.-D. Albakour, C. Macdonald, and I. Ounis. Experiments with a venue-centric model for personalised and time-aware venue suggestion. In *CIKM 2015: 24th ACM International Conference on Information and Knowledge Management*, 2015.

[24] R. Deveaud, M.-D. Albakour, J. Manotumruksa, C. Macdonald, I. Ounis, et al. Smartvenues: Recommending popular and personalised venues in a city. In *Proceedings of the 23rd ACM International Conference on Conference on Information and Knowledge Management*, pages 2078–2080. ACM, 2014.

[25] A. Dufaux, L. Besacier, M. Ansorge, and F. Pellandini. Automatic sound detection and recognition for noisy environment. In *Proc. of EUSIPCO'00.*

[26] M. D. Ekstrand, J. T. Riedl, and J. A. Konstan. Collaborative filtering recommender systems. *Foundations and Trends in Human-Computer Interaction*, 4(2):81–173, 2011.

[27] F. Grubb. Procedures for detecting outlying observations in samples. technometrics. 11, 1969.

[28] D. Guinard and V. Trifa. Towards the web of things: Web mashups for embedded devices. In *Proceedings of WWW '09*, 2009.

[29] V. Guralnik and J. Srivastava. Event detection from time series data. In *Proceedings of SIGKDD'99*, pages 33–42, 1999.

[30] S. Hansell. Google keeps tweaking its search engine. *New York Times*, June 2007. http://www.nytimes.com/2007/06/03/business/yourmoney/03 google.html

[31] S. H. Hashemi, M. Dehghani, and J. Kamps. Univ of Amsterdam at TREC 2015: Contextual suggestion track. In *Proceedings of TREC*, 2015.

[32] J. L. Herlocker, J. A. Konstan, A. Borchers, and J. Riedl. An algorithmic framework for performing collaborative filtering. In *Proceedings of the 22nd annual international ACM SIGIR conference on Research and development in information retrieval*, pages 230–237. ACM, 1999.

[33] D. Hiemstra, S. Robertson, and H. Zaragoza. Parsimonious language models for information retrieval. In *Proc. of SIGIR*, pages 178–185. ACM, 2004.

[34] L. Horváth. The maximum likelihood method for testing changes in the parameters of normal observations. *The Annals of statistics*, 21(2):671–680, 1993.

[35] R. Killick, P. Fearnhead, and I. Eckley. Optimal detection of changepoints with a linear computational cost. *arXiv*, 2011.

[36] Y. Koren, R. Bell, and C. Volinsky. Matrix factorization techniques for recommender systems. *Computer*, (8):30–37, 2009.

[37] X. Li, G. Cong, X.-L. Li, T.-A. N. Pham, and S. Krishnaswamy. Rank-geofm: A ranking based geographical factorization method for point of interest recommendation. In *Proceedings of the 38th International ACM SIGIR Conference on Research and Development in Information Retrieval*, pages 433–442. ACM, 2015.

[38] K. H. Lim. Recommending tours and places-of-interest based on user interests from geo-tagged photos. In *Proceedings of the 2015 ACM SIGMOD on PhD Symposium*, pages 33–38. ACM, 2015.

[39] T.-Y. Liu. Learning to rank for information retrieval. *Foundations and Trends in Information Retrieval*, 3(3):225–331, Mar. 2009.

[40] H. Ma, I. King, and M. R. Lyu. Learning to recommend with social trust ensemble. In *Proceedings of the 32nd international ACM SIGIR conference on Research and development in information retrieval*, pages 203–210. ACM, 2009.

[41] C. Macdonald and I. Ounis. Voting for candidates: adapting data fusion techniques for an expert search task. In *Proceedings of the 15th ACM international conference on Information and knowledge*

management, CIKM '06, pages 387–396, New York, NY, USA, 2006. ACM.

[42] J. Manotumruksa, C. Macdonald, and I. Ounis. Predicting contextually appropriate venues in location-based social networks. In Proceedings of CLEF 2016.

[43] A. K. McCallum. Mallet: A machine learning for language toolkit. http://mallet.cs.umass.edu, 2002.

[44] R. McCreadie, S. Vargas, C. Macdonald, I. Ounis, S. Mackie, J. Manotumruksa, and G. McDonald. Univ of Glasgow at TREC 2015: Experiments with Terrier in contextual suggestion, temporal summarisation and dynamic domain tracks. In *Proc. of TREC*, 2015.

[45] D. Metzler, C. Cai, and E. H. Hovy. Structured event retrieval over microblog archives. In *Human Language Technologies: Conference of the North American Chapter of the Association of Computational Linguistics, Proceedings*, June 3–8, 2012, Montréal, *Canada*, pages 646–655, 2012.

[46] D. O'Byrne, R. Brennan, and D. O'Sullivan. Implementing the draft w3c semantic sensor network ontology. In *Proceedings of PERCOM Workshops*, pages 196–201, 2010.

[47] M. Osborne, S. Petrovic, R. McCreadie, C. Macdonald, and I. Ounis. Bieber no more: First story detection using twitter and wikipedia. In *Proceedings of the SIGIR'12 Workshop on Time-aware Information Access. (TAIA).*, 2012.

[48] I. Ounis, G. Amati, V. Plachouras, B. He, C. Macdonald, and C. Lioma. Terrier: A High Performance and Scalable Information Retrieval Platform. In *Proceedings of ACM SIGIR-OSIR 2006*, Seattle, Washington, USA, 2006.

[49] P. Over, G. Awad, M. Michel, J. Fiscus, G. Sanders, W. Kraaij, A. F. Smeaton, and G. Quénot. TRECVID 2014 – An overview of the goals, tasks, data, evaluation mechanisms and metrics. In *Proc. of TRECVID'14.*

[50] S. Phan, T. D. Ngo, V. Lam, S. Tran, D.-D. Le, D. A. Duong, and S. Satoh. Multimedia event detection using segment-based approach for motion feature. *Journal of Signal Processing Systems*, 74(1):19–31, 2014.

[51] Y. Ren, M. Tomko, F. D. Salim, K. Ong, and M. Sanderson. Analyzing web behavior in indoor retail spaces. *Journal of the Association for Information Science and Technology*, 2015.

[52] S. Rendle. Factorization machines with libfm. *ACM Transactions on Intelligent Systems and Technology (TIST)*, 3(3):57, 2012.

[53] S. Robertson and H. Zaragoza. The probabilistic relevance framework: Bm25 and beyond. *Information Retrieval*, 3(4):333–389, 2009.

[54] T. Sakaki, M. Okazaki, and Y. Matsuo. Earthquake shakes twitter users: real-time event detection by social sensors. In *Proceedings of the 19th international conference on World wide web*, pages 851–860. ACM, 2010.

[55] J. Sankaranarayanan, H. Samet, B. E. Teitler, M. D. Lieberman, and J. Sperling. Twitterstand: news in tweets. In *Proceedings of GIS'09*, pages 42–51, New York, NY, USA, 2009. ACM.

[56] Y. Shi, M. Larson, and A. Hanjalic. List-wise learning to rank with matrix factorization for collaborative filtering. In *Proc. of RecSys*, pages 269–272. ACM, 2010.

[57] A. Singhal. Modern information retrieval: A brief overview. *IEEE Data Eng. Bull.*, 24(4):35–43, 2001.

[58] G. Sterling. Study: 43 percent of total google search queries sre local. http://searchengineland.com/study-43-percent-of-total-google-search-queries-have-local-intent-135428. Accessed: 5 October 2012.

[59] E. M. Voorhees. Evaluation by highly relevant documents. In *Proc. of SIGIR'01*, pages 74–82, 2001.

[60] Q. Wu, C. J. C. Burges, K. M. Svore, and J. Gao. Ranking, Boosting, and Model Adaptation. Technical Report MSR-TR-2008-109, Microsoft, 2008.

[61] D. Yang, D. Zhang, Z. Yu, and Z. Wang. A sentiment-enhanced personalized location recommendation system. In Proceedings of the 24th ACM *Conference on Hypertext and Social Media*, pages 119–128. ACM, 2013.

[62] M. Ye, P. Yin, and W.-C. Lee. Location recommendation for location-based social networks. In *Proceedings of the 18th SIGSPATIAL International Conference on Advances in Geographic Information Systems*, pages 458–461. ACM, 2010.

[63] Q. Yuan, G. Cong, Z. Ma, A. Sun, and N. M. Thalmann. Time-aware point-of-interest recommendation. In *Proceedings of the 36th international ACM SIGIR conference on Research and development in information retrieval*, pages 363–372. ACM, 2013.

[64] H. Zaragoza, B. B. Cambazoglu, and R. Baeza-Yates. Web search solved?: All result rankings the same? In *Proceedings of the 19th ACM International Conference on Information and Knowledge Management*, CIKM '10, pages 529–538, New York, NY, USA, 2010. ACM.

[65] J. Zhang and P. Pu. A recursive prediction algorithm for collaborative filtering recommender systems. In *Proceedings of the 2007 ACM conference on Recommender systems*, pages 57–64. ACM, 2007.

[66] A. Zubiaga, D. Spina, V. Fresno, and R. Martínez. Classifying trending topics: a typology of conversation triggers on twitter. In *Proceedings of CIKM'11*, pages 2461–2464, New York, NY, USA, 2011. ACM.

4

Development Tools for IoT Analytics Applications

John Soldatos and Katerina Roukounaki

Athens Information Technology, Greece

4.1 Introduction

The proliferation of IoT analytics applications has recently created a need for tools and techniques that could support developers in the task of producing and deploying IoT analytics services. In principle, IoT analytics development tasks can be supported by readily available tools for IoT applications and their combination with tools and techniques for data mining and analytics. In particular, IoT development tools undertake to collect and appropriately pre-process data streams stemming from IoT systems (including "live" high-velocity data streams), while conventional data analytics tools can be used to analyze the information contained in these data streams towards extracting knowledge. Hence this combination brings together the IoT and BigData worlds, thus facilitating developers in the task of implementing and deploying IoT analytics applications.

This chapter is destined to present this blending of IoT development tools and data analytics tools. In particular, the chapter is devoted to the presentation of sample tools for the development of IoT analytics applications and more specifically the development tools of an IoT platform which has been recently developed as part of the FP7 VITAL project. These tools support IoT development functionalities such as discovery of data streams from IoT systems, filtering of data streams in order to economize on bandwidth and storage resources, as well as semantic unification of heterogeneous streams in order to facilitate the unified processing of diverse data sources. The importance of such functionalities for IoT analytics applications is adequately described in the scope of other chapters of this book, along with specific technology solutions for their implementation. The present chapter considers these functionalities as part of the presented development tools infrastructure, which leverages middleware services (e.g., data streams discovery and filtering) of the VITAL platform. Therefore, the chapter introduces these middleware services as well, along with their positioning in the overall architecture of the VITAL platform.

The VITAL development tools are based on the popular Node-RED tool for IoT applications, which has been customized to the needs of the VITAL platform. The customization of the Node-RED tool included also the enhancement of data mining and data analytics functionalities, which are illustrated in the scope of this chapter. Along with the VITAL development tools, the VITAL platform provides also a tool for managing IoT resources (including IoT data sources and data streams), including configuration, security and SLA (Service Level Agreement) management functionalities. The latter can greatly facilitate the monitoring of IoT analytics applications and can be used in conjunction with the VITAL development tools. Therefore, we also present the VITAL development tools as an integral element of the wider suite of tools that support developers in the production of IoT analytics applications. The development and management tools are bundled in an integrated development environment, which is accessible over the web and from a single entry point.

Overall, the chapter is structured as follows: The next paragraph discusses relevant work on development tools for IoT analytics. Following chapters illustrate the VITAL architecture and the middleware services that are used in order to support the functionalities of the tools in the scope of the VITAL platform. Along with these functionalities, the chapter discusses the VITAL development tools with particular emphasis on their add-on features which enhance Node-RED. The discussion includes also insights on the limitations of the development tools, which could be remedied as part of future work. Moreover a dedicated section is devoted to the description of the VITAL management environment. Indicative applications are finally presented in order to illustrate the added-value of the tools and the productivity boost that they can offer to large number of developers of IoT analytics applications.

4.2 Related Work

The provision of development environments for IoT analytics has its roots on tools and techniques for the development for IoT applications and data analytics. IoT development tools provide the means for interfacing to IoT systems towards collecting, filtering and fusing IoT data streams.

At the same time, data analytics environments provide the means for developing and executing data analytics algorithms.

Early IoT development tools have been introduced as part of WSN (Wireless Sensor Network) platforms (e.g. [1, 2]) and RFID (Radio-frequency Identification) platforms (e.g. [3]). Recently we have witnessed the emergence of integrated development environments and tools for wider classes of IoT applications, including visual modelling tools following Model Driven Architectures (MDA) (e.g. [4, 5]).

There have also been IoT development environments associated with mainstream IDE projects, such as Eclipse Kura, which is an Eclipse IoT project that provides a framework for M2M service gateways (i.e., devices that act as mediators in the machine-to-machine and the machine-to-Cloud communication). Kura facilitates the development, deployment and remote management of M2M applications and its use requires only the installation of an Eclipse plugin on the developer's machine. It is based on Java and OSGi, the dynamic module system for Java, and it can be used to turn a Raspberry Pi or a BeagleBone Black into an IoT gateway.

Node-RED is another open-source project that is focused on IoT. This project is reused and extended as part of the prototype implementation presented in this chapter. It is described in the following paragraph in order to facilitate the understanding of the approach and the related implementation.

Integrated Cloud Environments (ICEs) have come to change this workflow, by turning development environments from products into services. ICEs are essentially IDEs that are usually web accessible, and that leverage the Cloud into the software development lifecycle. In order to use an ICE, developers do not need to install any more tools on their machines; all they need to do is log into a web site (that acts as the entry point to the ICE), and start using it. In this case, most of the tasks take place in the Cloud; some ICEs use the Cloud even to store the developers' code.

While IoT tools provide the means for interfacing to data sources towards accessing, processing and combining data streams, they do not typically offer capabilities for analyzing IoT data. Therefore, their use for IoT analytics requires their integration with data analytics libraries and tools such as:

- The Technical Analysis library (http://ta-lib.org/), which is an open source library that enables technical analysis of financial markets data.
- The Java Universal Network Graph (http://jung.sourceforge.net/), which enables the analysis and visualization of graph or network based data (e.g., social networks data).

- The GeoTools (http://www.geotools.org/) toolkit, which enables the manipulation of GIS data, including the analysis of their spatial and non-spatial attributes or GIS data.
- The R project (https://www.r-project.org/), a highly extensible environment, which enables the execution of a wide variety of statistical (e.g., linear and nonlinear modelling, classical statistical tests, time-series analysis, classification, clustering) and graphical techniques.

The scope of the work that is presented in following paragraphs, involves the integration of the R project within an enhanced version of the Node-RED tool, as part of an integrated development environment offer by the VITAL smart cities platform (developed in the scope of the FP7 VITAL project). Note that the integration of IoT tools with data analytics tools is also evident in the scope of popular public cloud environment (such as the Amazon EC2 and the Microsoft Azure cloud services), which provide functionalities for IoT applications development along with data analytics toolkits.

4.3 The VITAL Architecture for IoT Analytics Applications

The VITAL IoT development environment is an integral part of the VITAL smart cities platform. This platform provide a range of tools and techniques for developing, deploying, managing and operating IoT applications in smart cities, including applications that leverage data and services from multiple IoT systems and data sources. The latter applications are based on the semantic interoperability features of the VITAL platform, which enable the repurposing and reuse of services and datasets from multiple IoT systems. An overview of the VITAL platform is provided in Figure 4.1.

The main components of the platform are:

- **Platform Provider Interface (PPI)**: The PPI is an abstract interface to underlying IoT systems and data sources, including the large number of legacy IoT systems that are nowadays available in the scope of digitally mature cities. PPI provides access to both metadata and data of the underlying systems, In particular, the information that is specified in the PPI covers system-level information, information about the internet-connected objects of the system, sensors-based observations' information (data), as well as metadata for managing SLAs (Service Level Agreements) between the operator of the VITAL platform (e.g., city authorities, telecom services providers) and the operators of the individual IoT systems.

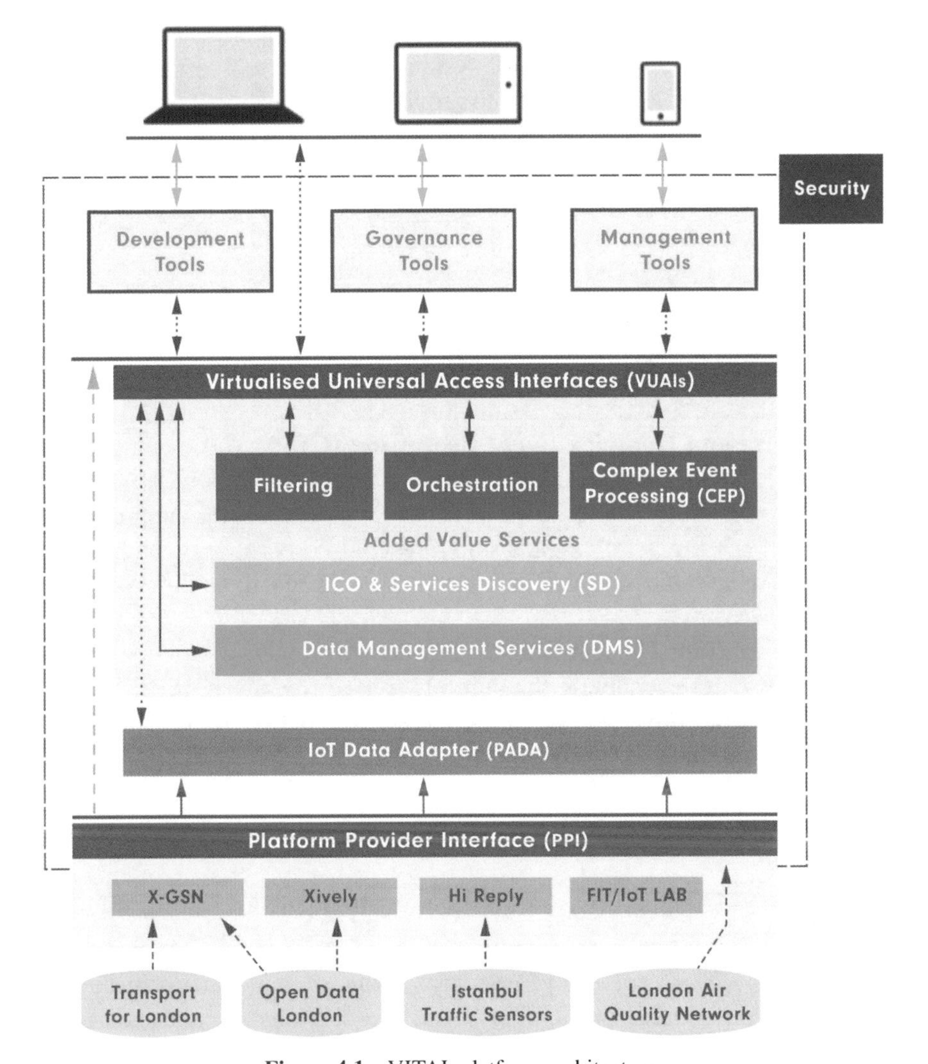

Figure 4.1 VITAL platform architecture.

- **Data Management Service (DMS)**: This is a data service (empowered by scalable operational databases), which persists and manages data from all of the underlying IoT systems. Data within the DMS are semantically unified, since they comply with the same data model (schema, ontology). Note that the DMS provides interoperable cached data from the various IoT systems, thus providing a foundation for the provision of a range of Data-as-a-Service (DaaS) services.

- **IoT Data Adapter (PADA)**: This component manages the subscriptions of the VITAL platform to IoT systems and data sources, through the management of PPIs. It therefore provides functionalities for registering and deregistering PPIs as data contributors to the DMS, while at the same time managing data acquisition from the IoT systems to the DMS (according to a publish-subscribe paradigm).
- **IoT Service Discovery (SD)**: This component enables the discovery of services, sensors, internet-connected devices and other IoT resources. In the SD context, the term "services" refers to services provided by the VITAL platform (possibly assembled based on the orchestrator component) rather than to low-level services provided by the IoT systems. The latter are typically accessible through PPIs.
- **Filtering and Complex Event Processing (CEP)**: These components offer data filtering and event generation functionalities based on data streams residing with the DMS. The filtering components support static data processing, with emphasis on threshold-based filtering and resampling. At the same time, CEP supports both static and dynamic processing of IoT streams.
- **Orchestration**: This component provides functionalities for composing workflows, thus enabling the orchestration of (composite) IoT services based on more elementary ones. As already outlined, composite IoT services produced by the orchestrator are registered to the SD component.
- **VUAIs (Virtualized Unified Access Interfaces)**: These are interfaces enabling IoT system agnostic access to data and services of the VITAL platform.

On top of the VITAL platform, three distinct environments are offered, namely:

- A management environment providing FCAPS (Fault Configuration Accounting Performance and Security) management functionalities for the VITAL modules, but also for the data and services from the underlying IoT systems.
- A governance environment enabling the configuration of the VITAL platform (including configuration of its individual modules) according to the needs and characteristics of a given urban environment. The governance environment takes into account information and parameters such as the geography and the demographics of the city in order to appropriately customize the operation of the VITAL platform.
- A development environment for producing smart city applications based on the VITAL platform. It extends the popular Node-RED tool on the

basis of functionalities for the VITAL modules, thus enabling developers to combine VITAL functionalities (e.g., orchestration, filtering, semantic interoperability) with the rich set of Node-RED functionalities. It also integrates the R project in order to boost the development of IoT Analytics applications.

Following paragraphs illustrate the VITAL development environment, as a concrete example of a tool that facilitates the development of IoT Analytics applications.

4.4 VITAL Development Environment

4.4.1 Overview

The primary goal of the VITAL development environment is to integrate all functionalities provided by the VITAL platform and make them accessible to smart city application developers through a single tool, the VITAL development tool. To this end, the various functionalities of the VITAL platform are integrated into the tool based on VUAIs, which are currently implemented as RESTful web services. This renders Node-RED ideal as the basis for the implementation of the VITAL development tool. Furthermore, the growing number of nodes (i.e. development functionalities) that are available for Node-RED, as well as its simplicity, user-friendliness, extensibility and popularity led to the selection of Node-RED as a basis for developing the VITAL tool. As shown in Figure 4.2, the VITAL development tool is based on the enhancement of Node-RED with a number of VITAL-related nodes, as well as with functionalities provided by the R project (and associated programming language for statistical computing and graphics). The result of this enhancement process is an easy-to-use tool that also enables its users to perform a number of VITAL-related (e.g. retrieval of IoT system metadata) and data analysis (e.g. data value prediction or data clustering) tasks, based on the exploitation of the VITAL platform.

A short overview of the extra nodes that have been added to the core node palette of Node-RED for the purpose of supporting and exposing the VITAL functionalities follows.

4.4.2 VITAL Nodes

In order to expose the functionalities provided by the VITAL platform through the VITAL development tool, a number of new Node-RED nodes

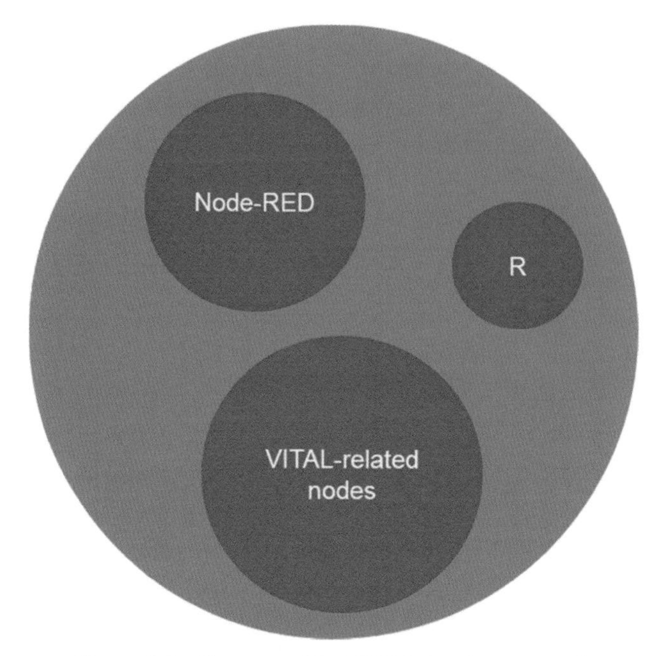

Figure 4.2 Elements of the VITAL development tool.

were created and added to the tool. More specifically, the node palette was complemented with the following node categories: (1) **ppi** that contains nodes to use in order to communicate directly with PPI-compliant IoT systems and data sources, (2) **data** that contains nodes that expose the functionalities provided by the DMS, (3) **discovery** that contains nodes that enable the discovery of different types of IoT resources, and (4) **filtering** that contains nodes that expose the filtering functionalities provided by the VITAL platform.

4.4.2.1 PPI nodes

Each node in the **ppi** category corresponds to a primitive specified as part of the Platform Provider Interface.

4.4.2.2 System nodes

System nodes are used to retrieve metadata about a PPI-compliant IoT system. When a system node receives a message, the node accesses the relevant primitive of the PPI implementation exposed by that system, and puts the result (i.e., the system metadata) into the message it finally sends out.

4.4.2.3 Services nodes

Services nodes retrieve metadata about the IoT services that a PPI-compliant IoT system provides. The message that a services node receives may contain information, which is used to filter the services to retrieve metadata for (based on their **ID** and **type**), whereas the message that a services node sends contains the retrieved service metadata.

4.4.2.4 Sensors nodes

Sensors nodes are function nodes that retrieve metadata about sensors that an IoT system manages. The messages sent to these nodes can be used to filter the sensors to retrieve metadata for (based on their **ID** and **type**), whereas the messages sent by these nodes contain the retrieved sensor metadata.

4.4.2.5 Observations nodes

Observations nodes are function nodes that pull observations made by sensors managed by a PPI-compliant IoT system. Input messages may contain information, which can be used to filter the observations to fetch (based on the **sensor** that made them, the observed **property** and the **time** when they were made), whereas output messages contain the retrieved observations.

4.4.2.6 DMS nodes

Nodes that expose functionalities provided by the DMS component of the VITAL platform fall into the **data** category.

4.4.2.7 Query systems

Query systems nodes query DMS for systems that meet specific criteria. The message that a query systems node receives contains a **query**, whereas the message that it sends out contains the metadata about all IoT systems that are registered with the VITAL platform and match the query.

4.4.2.8 Query services

Query services nodes are used to retrieve information about IoT services based on specific criteria. Input messages contain **queries**, whereas output messages contain metadata about IoT services that match those queries.

4.4.2.9 Query sensors

Query sensors nodes query DMS for internet-connected objects that meet specific criteria. The messages sent to these nodes contain a **query**, whereas the messages that these nodes send as a response contain metadata about all internet-connected objects that match the given query.

4.4.2.10 Query observations

Query observations nodes query DMS for observations. The message that a query observations node receives contains a **query**, whereas the message that a query observations node sends contains observations based on the given query.

4.4.2.11 Discovery nodes

The **discovery** node category groups together all Node-RED nodes that enable the discovery of different types of IoT resources by leveraging the discovery functionalities provided by the VITAL platform.

4.4.2.12 Discover systems nodes

Discover systems nodes are used to discover systems based on their **type** and/or **spatial context**. The messages sent to discover systems nodes contain the criteria, whereas the messages sent by these nodes contain the metadata about the systems that meet these criteria.

4.4.2.13 Discover services nodes

Discover services nodes enable the discovery of services based on specific criteria. Input messages may contain a **type** and a **system** URI, and output messages contain the available metadata about all services of that type that are provided by that system.

4.4.2.14 Discover sensors nodes

Discover sensors nodes are used to discover sensors based on their **position** (current or within a specified **time window**), **type**, **movement pattern**, **connection stability**, and whether they provide a **localizer service**. Input messages contain the criteria that sensors must meet, whereas output messages contain metadata about the sensors that meet them.

4.4.2.15 Filtering nodes

Filtering nodes are used to access the VITAL filtering functionalities.

4.4.2.16 Threshold nodes

Threshold nodes perform threshold-based filtering to the values collected from a specific internet-connected object, for a specific property, in a specific area, and within a specific time interval. Messages sent to threshold nodes contain **criteria**, based on which to retrieve observations, a **threshold** value, and a **relation**, and messages sent by these nodes contain all values that meet the specified criteria and have the specified relation with the specified threshold.

4.4.2.17 Resample nodes

Resample nodes are used to resample (down-sample or up-sample) data streams using a different time interval than the one they were initially sampled with. Input messages specify the **data stream** (i.e., the **sensor** and the observed **property**), the new **time interval**, and the **time period**, over which to perform the resampling, whereas output messages contain the resampled observations.

4.5 Development Examples

4.5.1 Example #1: Predict the Footfall!

The purpose is to implement a web page that shows a map of Camden town. When the user clicks anywhere on that map, a pop-up appears that informs the user about the people that are expected to be walking around that area during the next hour. The expected result is shown in Figure 4.3.

In order to provide the required functionality, two flows were created using the VITAL development tool. The first flow is a web service that responds with the static HTML page that contains the Camden map. The second flow is a web service that given a location responds with a prediction for the number of people in that area within the next hour. Both flows are depicted in Figure 4.4.

The second flow receives a location, uses a **query sensors** node to find the footfall sensor that is closer to that location, uses an **observations** node to retrieve observations collected from that sensor in the last ten days, and finally leverages the **rstats** package to predict the value of that sensor in the next hour.

4.5.2 Example #2: Find a Bike!

The purpose is to build a web page that people that move in London can use in order to find out whether there are any bikes available near them. The user specifies their location on the map, and as a result a marker appears on the map for each docking station within a 500 m radius that has at least one available bike. Figure 4.6 shows the implemented web page.

Figure 4.7 depicts the two flows that were implemented for the purposes of this example. The first flow implements the web service that returns the static HTML page. The second flow receives the current location of the

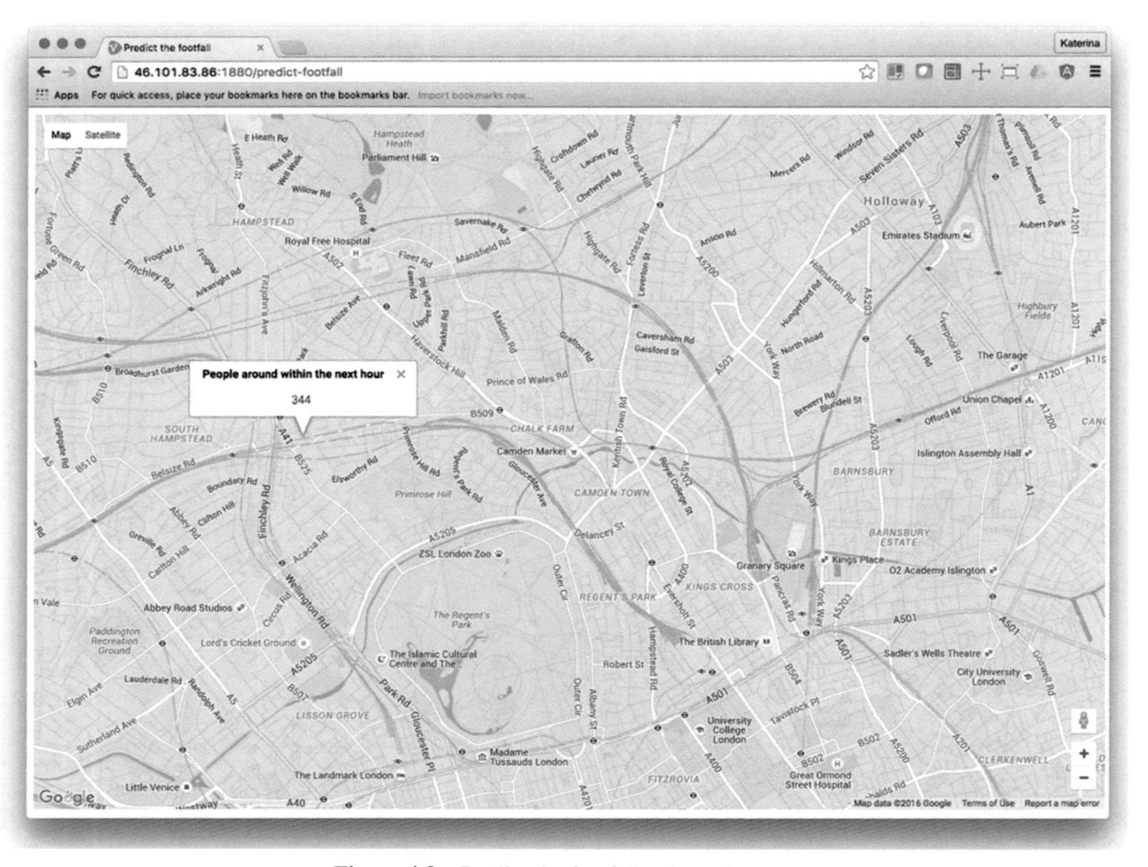

Figure 4.3 Predict the footfall – the web page.

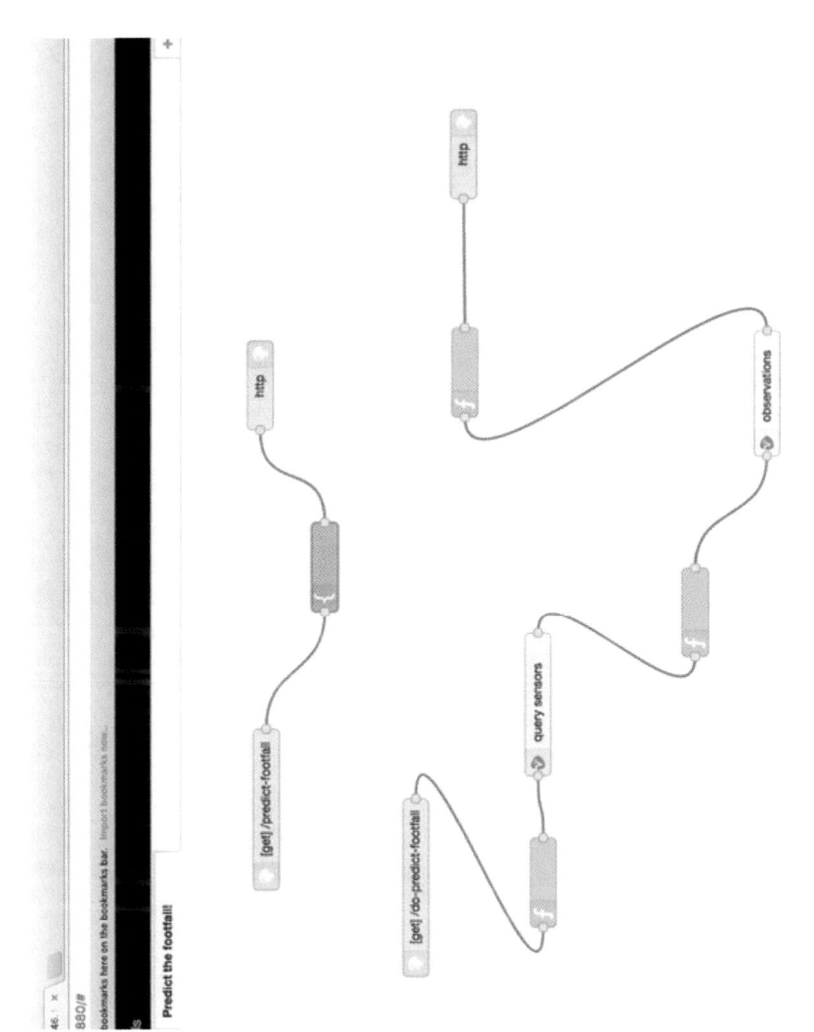

Figure 4.4 Predict the footfall – the flows.

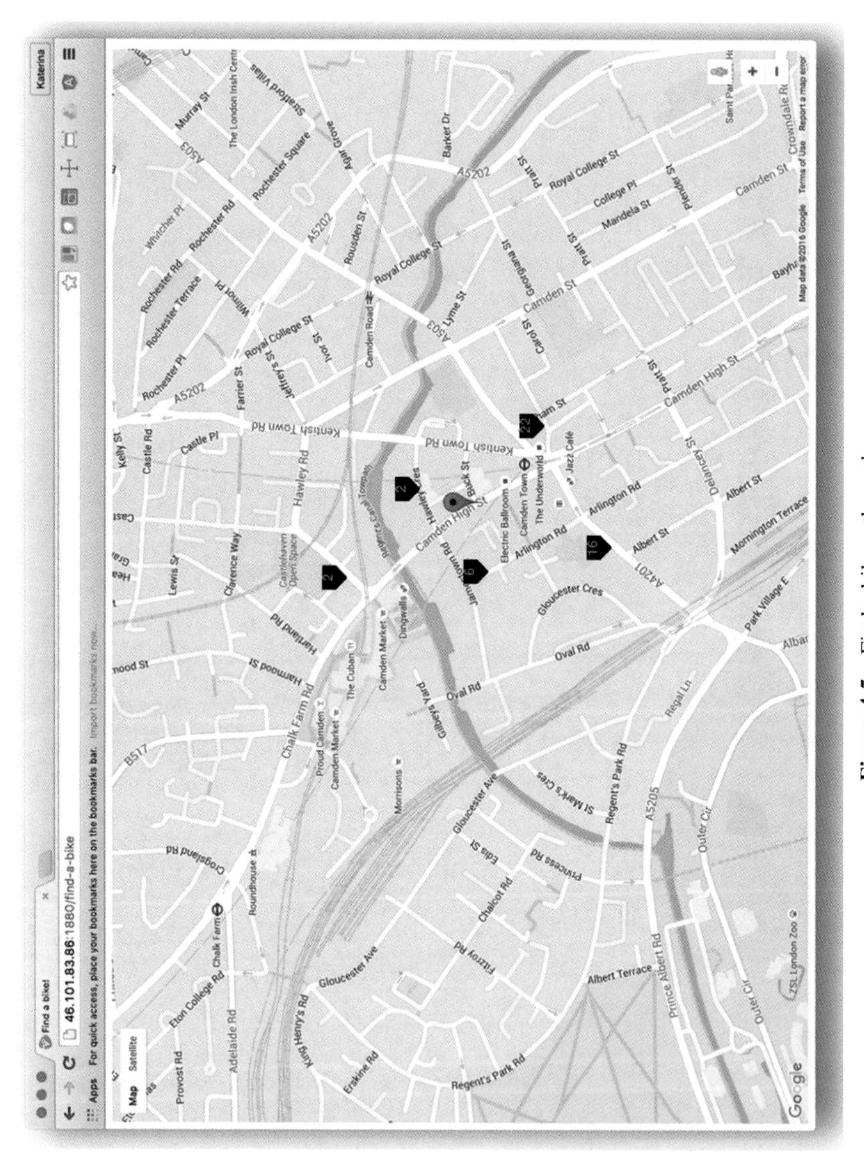

Figure 4.5 Find a bike – the web page.

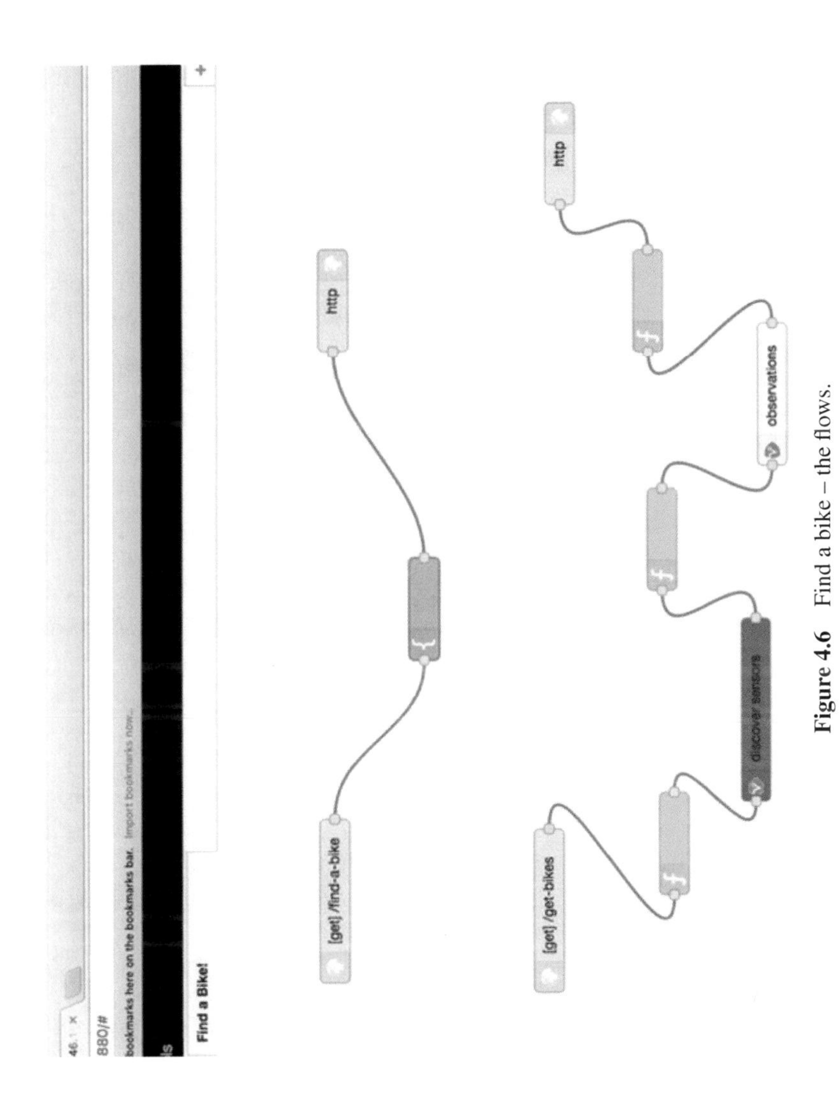

Figure 4.6 Find a bike – the flows.

user (that they have specified by clicking on the map), discovers all docking stations in that area (using a **discover sensors** node, since docking stations are essentially sensors), finds out how many available bikes each one of these stations has (using an **observations** node, since this implies the retrieval of the last observation made by each one of the corresponding sensors), and finally responds with the locations of the stations that have at least one available bike.

4.6 Conclusions

As IoT analytics applications, proliferate developers are starving for tools that can boost their development productivity. The wide array of emerging tools for IoT and data analytics applications are not enough to maximize developers' productivity, when used in isolation. Their combination and integration is therefore needed in order to achieve multiplicative benefits, i.e. leverage productivity benefits from both analytics and IoT tools. Moreover, in several cases the integration of data streaming concepts is also important, given the high velocity of IoT data streams. Integration of data streaming tools was not extensively presented in the scope of the Chapter, as VITAL stores IoT data into a scalable datastore in a semantically unified manner. However, the presented approach demonstrates also the merits of semantic interoperability for the development of added-value IoT analytics applications in smart cities, notably applications that leverage and process data from multiple IoT systems and data sources, which have typically been developed and deployed independently.

Acknowledgements

Part of this work has been carried out in the scope of the VITAL project (www.vital-iot.eu), which is co-funded by the European Commission in the scope of the FP7 framework programme (contract No. 608662).

References

[1] Karl Aberer, Manfred Hauswirth, Ali Salehi. *Infrastructure for Data Processing in Large-Scale Interconnected Sensor Networks*. MDM 2007: 198–205.

[2] Ioannis Chatzigiannakis, Georgios Mylonas, Sotiris E. Nikoletseas. *50 ways to build your application: A survey of middleware and systems for Wireless Sensor Networks*. ETFA 2007: 466–473.

[3] Achilleas Anagnostopoulos, John Soldatos, Sotiris G. Michalakos. REFiLL. *A lightweight programmable middleware platform for cost effective RFID application development.* Pervasive and Mobile Computing 5(1): 49–63 (2009).

[4] P. Patel, A. Pathak, T. Teixeira, and V. Issarny. *Towards application development for the internet of things.* In Proceedings of the 8th Middleware Doctoral Symposium, page 5. ACM, 2011.

[5] Pankesh Patel, Animesh Pathak, Damien Cassou, Valérie Issarny. *Enabling High-Level Application Development in the Internet of Things*, Lecture Notes of Sensor Systems and Software, the Institute for Computer Sciences, Social Informatics and Telecommunications Engineering Volume 122, 2013, pp. 111–126.

[6] D. Cassou, J. Bruneau, J. Mercadal, Q. Enard, E. Balland, N. Loriant, and C. Consel. *Towards a tool-based development methodology for sense/compute/control applications.* In Proceedings of the ACM international conference companion on Object oriented programming systems languages and applications companion, pages 247–248. ACM, 2010.

[7] L. Atzori, A. Iera, and G. Morabito. *The internet of things: A survey.* Computer Networks, 54(15): 2787–2805, 2010.

[8] J. S. Brown and R.R. Burton. Diagnostic models for procedural bugs in basic mathematical skills. *Cognitive Science*, 2(2):155–192, 1978.

[9] A. Cawsey. Planning interactive explanations. I*nternational Journal of Man-Machine Studies*, in press.

5

An Open Source Framework for IoT Analytics as a Service

John Soldatos[1], Nikos Kefalakis[1] and Martin Serrano[2]

[1]Athens Information Technology, Greece
[2]Insight Center for Data Analytics, National University of Ireland, Galway, Ireland

5.1 Introduction

Earlier chapters have illustrated the importance of IoT and cloud computing convergence, as a means of achieving scalability and meeting QoS (Quality of Service) constraints. IoT deployments in the cloud are motivated by two main business drivers:

- **Business Agility**: Cloud computing alleviates tedious IT procurement processes, since it facilitates flexible, timely and on-demand access to computing resources (i.e. compute cycles, storage) as needed to meet business targets. In the case of IoT analytics applications, IoT developers and deployments can flexibly gain access to the storage and processing resources that they need in order to support their applications.
- **Reduced Capital Expenses**: Cloud computing leads to reduced capital expenses (CAPEX) (i.e. IT capital investments), through converting CAPEX to operational expenses (OPEX) (i.e. paying per month, per user for each service). This is due to the fact that cloud computing enables flexible planning and elastic provisioning of resources instead of upfront overprovisioning. Among the benefits of such flexibility is that it enables small and medium size enterprises (SMEs) to adopt a pay-as-you-go and pay-as-you-grow model to infrastructure acquisition and use, through paying for the computing resources and capacity that they need. This can be particularly important for the proliferating number of SMEs (including

high-tech startups), which exploit IoT analytics as part of their products or services.

Similarly to cloud computing infrastructures [1], integrated IoT/cloud infrastructures and related services can be classified to the following models:

- **Infrastructure-as-a-Service (IaaS) IoT/Clouds**: These services provide the means for accessing sensors and actuator in the cloud. The associated business model involves the IoT/Cloud provide to act either as data or sensor provider. IaaS services for IoT provide access control to resources as a prerequisite for the offering of related pay-as-you-go services.
- **Platform-as-a-Service (PaaS) IoT/Clouds**: This is the most widespread model for IoT/cloud services, given that it is the model provided by all public IoT/cloud infrastructures outlined above. As already illustrate most public IoT clouds come with a range of tools and related environments for applications development and deployment in a cloud environment. A main characteristic of PaaS IoT services is that they provide access to data, not to hardware. This is a clear differentiator comparing to IaaS IoT clouds.
- **Software-as-a-Service (SaaS) IoT/Clouds**: SaaS IoT services are the ones enabling their uses to access complete IoT-based software applications through the cloud, on-demand and in a pay-as-you-go fashion. As soon as sensors and IoT devices are not visible, SaaS IoT applications resemble very much conventional cloud-based SaaS applications. There are however cases where the IoT dimension is strong and evident, such as applications involving selection of sensors and combination of data from the selected sensors in an integrated applications. Several of these applications are commonly called Sensing-as-a-Service, given that they provide on-demand access to the services of multiple sensors. Note that SaaS IoT applications are typically built over a PaaS infrastructure and enable utility based business models involving IoT software and services.

Although the Sensing-as-a-Service paradigm is a special case of an SaaS deployment, it is in practice applicable to IoT applications only. Indeed, Sensing-as-a-Service applications involve on-demand collection, processing and analysis of information from sensors (i.e. IoT devices) [2]. The on-demand and dynamic nature of Sensing-as-a-Service applications in reinforced by the location dependent and time dependent nature of such IoT applications, which permit the dynamic selection of the IoT resources (sensors) that will provide the data streams to be processed. As such Sensing-as-Service can be seen as a case of an "IoT Analytics as a service" paradigm, where the IoT application users is allowed to dynamically specified data processing and analytics functionalities, along with the IoT devices on which they will be executed.

In this chapter we present a framework for implementing Sensing-as-a-Service applications based on the open source OpenIoT project [3]. The OpenIoT framework enables the dynamic selection of sensors and resources, as well as the subsequent specification of processing functionalities over the data of the selected sensors. In essence it enables the specification of dynamic sensor queries, which can be considered the first step towards IoT analytics as a service [4]. In addition to facilitating the dynamic definition and deployment of such Sensing-as-Service (or IoT analytics as a service) services, OpenIoT provides:

- Semantic interoperability and unification of data from diverse IoT sensors and other data sources, through ensuring their conversion and compliance to a common ontology, namely the OpenIoT ontology, which is an extended version of the W3C SSN (Semantic Sensor Networks) ontology [5].
- A range of easy-to-use tools for the visual specification of the Sensing-as-a-Service services. The tools enable the definition and deployment of SPARQL based sensor queries, through exploiting sensors registered within the OpenIoT framework.

Note that OpenIoT does not provide sophisticated data analytics functionalities, but it can well be extended on the basis of frameworks for data mining and machine learning, in order to support more advanced analytics functionalities. Such extensions are worked out in the scope of the H2020 FIESTA-IoT project, which provide functionalities for semantically interoperable IoT experimentation i.e. the execution of data-centric IoT experiments based on data streams from multiple IoT experimental facilities. Following sections of the chapter focus on the description of the OpenIoT framework and capabilities for Sensing-as-a-Service, along with a practical example of constructing and deploying a relevant sensor query based on the OpenIoT tools. Moreover, the enhancement of the Sensing-as-a-Service paradigm with more sophisticated analytics functionalities, towards an IoT Analytics as a Service paradigm is also discussed.

5.2 Architecture for IoT Analytics-as-a-Service

5.2.1 Properties of Sensing-as-a-Service Infrastructure

Service formulation and delivery in the scope of OpenIoT is characterized by the following properties:

- **On-demand**: Service formulation and delivery in OpenIoT should be performed on-demand. This implies the need for on-demand expressing requests for IoT services formulation, which shall be fulfilled by the OpenIoT middleware infrastructure. Therefore, service formulation should provide the means to dynamically selecting sensors and ICOs needed in order to satisfy the demanded service requests.
- **Cloud-based**: OpenIoT services are provided in a cloud environment. At the heart of this environment lies a scalable sensor cloud infrastructure, which shall provide sensor data access services. Thus, the OpenIoT service formulation strategies must take into account the need to access, use and combine services residing within the sensor/ICO cloud.
- **Utility-based**: Service delivery in OpenIoT is utility-based, which is in-line with the on-demand and cloud-based properties. As a result, OpenIoT should provide the means for calculating utility, through making provisions for storing a range of utility parameters (e.g., usage parameters for the employed ICOs) during the process of service formulation.
- **Service-Oriented**: OpenIoT requests will result in the deployment of services. The latter may be the composition of other services, such as services for accessing data streams in the cloud. Overall, OpenIoT has a service-oriented nature.
- **Optimized**: OpenIoT incorporates a wide range of self-management and self-optimization algorithms. The service formulation process ensures that information about resources reservation and usage is recorded in order to enable the implementation of utility-based optimization algorithms.

5.2.2 Service Delivery Architecture

The architecture of the OpenIoT platform is illustrated in Figure 5.1, while a more detailed overview of the interactions between the various modules is depicted in Figure 5.2. As already outlined, OpenIoT enables the cloud-based delivery of IoT data processing services, through enabling the creation of dynamic on-demand services. These services select and process data from a multitude of different data sources.

Overall, the architecture makes provisions for the creation and fulfillment of requests for services to the OpenIoT system. It is empowered by the following components:

- **Service Request Definition Component ("Request Definition")**: Service Request Definition is the component where requests for IoT

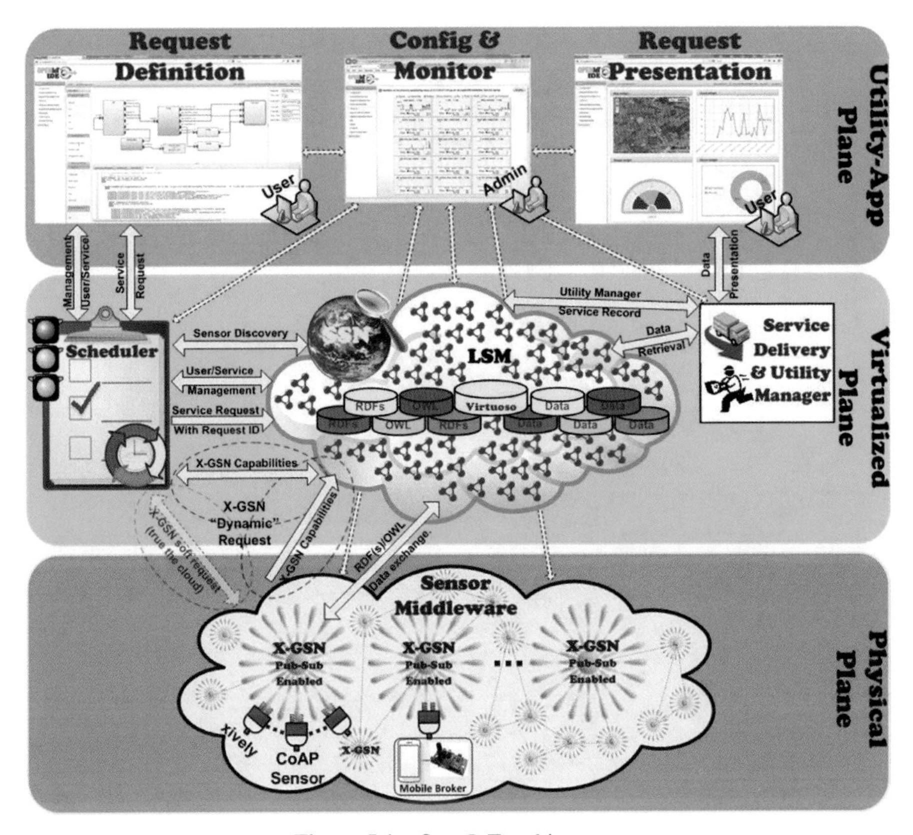

Figure 5.1 OpenIoT architecture.

services are formulated (by end-users) and accordingly submitted to the OpenIoT system. This component comes with an appropriate graphical user interface (GUI), which facilitates the service request customization.

- **(Global) Scheduler**: The Global Scheduler is in charge of accepting and prioritizing the various service requests (by one or more end-users) and accordingly generates the list of sensors (and other Internet Connected Objects (ICO)) that participate in the delivery of the service. Furthermore, the global Scheduler performs the required reservations of resources, which facilitate utility calculation and resource optimizations.
- **Service Discovery**: Service discovery refers to the OpenIoT directory services. It maintains the semantically annotated descriptions of the sensors that are known to the OpenIoT system. Service discovery

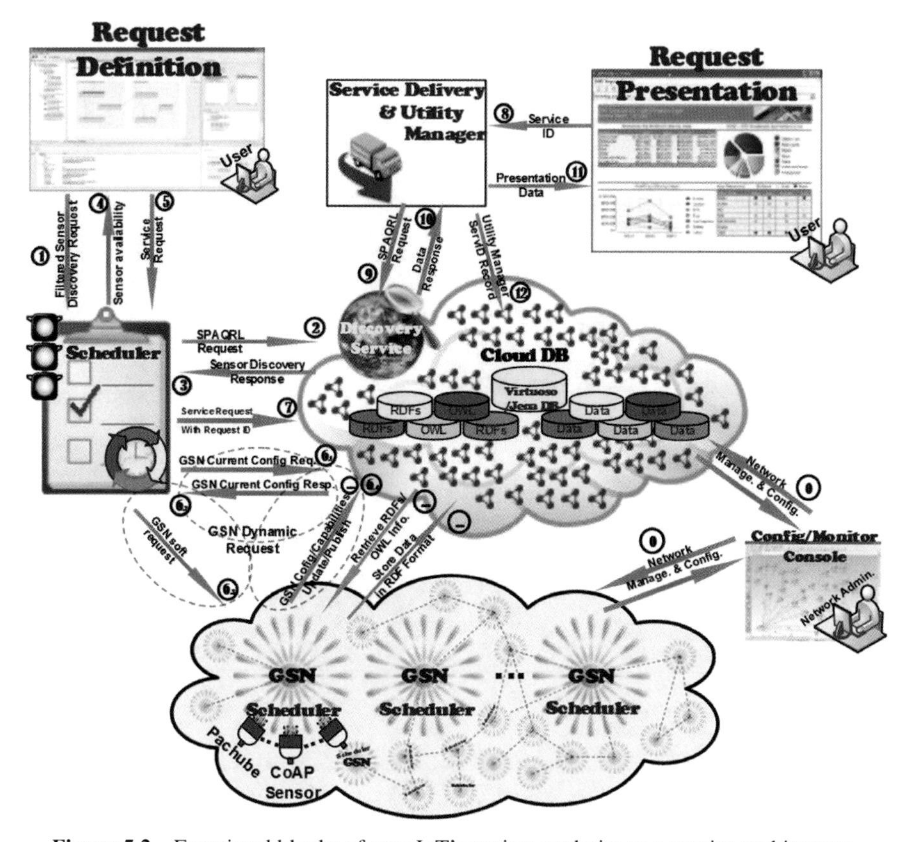

Figure 5.2 Functional blocks of openIoT's project analytics as a service architecture.

relies on the registration of sensors in the directory service repository. The structure of the service directory is based on the OpenIoT ontology (an enhanced version of the W3C SSN ontology).

- **Cloud Infrastructure**: This refers to the cloud computing environment (functional and operational), which ensures sensor cloud integration and streaming of sensors and ICO data to cloud storage, the operations performed in the cloud infrastructure are independent of the infrastructure management and the infrastructure modifications.
- **Global Sensor Networks (GSN) Nodes**: GSN nodes refer to deployment instances of the GSN middleware [6]. They play a significant role in the data provisioning for the IoT service delivery, since they enable the interfacing of physical-world devices to the OpenIoT system (via the cloud infrastructure). At the same time, GSN nodes perform a range of

local-level optimizations, on the basis of the ICOs that they comprise of and on how these participate and contribute to the various services, prioritizing and transforming the data acquired form the physical sensors into normalized data.

- **Service Delivery and Utility Manager (SD&UM)**: The service delivery manager ensures the proper assembly and delivery of the services subject to the various constraints imposed either for physical infrastructure restrictions or service customization definitions. To this end, it uses the selected sensors and ICOs and combines them as specified in the service request sent to the system. The combination depends also on the optimizations performed by the OpenIoT infrastructure, given that these optimizations may, for example, regulate the frequency of accesses to the various underlying data services.
- **Service Presentation ("Request Presentation")**: This component facilitates the implementation of the presentation layer of the service, on the basis of mashups and other visualization libraries. It can be considered as an optional component aiming at easing the presentation of the services according to the preferences and needs of the end-user.

5.2.3 Service Delivery Concept

In-line with the main components of the OpenIoT architecture outlined above, service delivery is based on the selection and orchestration of multiple services (including cloud services) that provide data and/or instigate tasking or actuation functionalities. The orchestration and combination of those services is based on the following factors:

- **Type of (requested) service**: The service request specifies different possible operations on ICOs, such as selection, retrieval and processing of their data, or execution of actuation commands. The OpenIoT sensor cloud infrastructure can be seen as a large-scale distributed sensors and ICO database. Service requests can be thought of as queries and operations over this database (i.e. SQL (Structured Query Language) can be thought as a representative metaphor). Depending on the query and operation, the OpenIoT infrastructure will instigate alternative paths within the service delivery strategies. For example, query operations (i.e. «SELECT» in SQL terms) will lead to the combination of sensor data access services, while actuating operations (i.e. «UPDATE» or «EXECUTE» operations) will lead to the invocation of the actuating services. Furthermore, requests combining both actuation and selection functions should trigger alternative paths within the OpenIoT service formulation strategies.

- **Optimizations**: The OpenIoT sensor cloud is a self-managing infrastructure, which provides opportunities for optimal delivery of the services. Therefore, the resource management and optimization capabilities of the OpenIoT infrastructure affect service formulation and delivery. Alternative service delivery and execution paths are likely to be considered in the scope of the optimization of the OpenIoT services.
- **Sensor and ICO selection**: The selected sensors and ICOs influence the formulation and delivery of services. Different ICOs may provide different capabilities in terms of data selection and actuating services execution. Therefore, the OpenIoT service delivery environment deals with the heterogeneity of data being collected from the various ICOs. In particular, the OpenIoT cloud and the underlying GSN nodes provide a virtualized interface for accessing the low-level capabilities of the ICOs acting as data collectors for the OpenIoT system.

Finally, the service formulation and delivery mechanisms consider the need to support both service deployment and service un-deployment. Service un-deployment should be implementing as an integral element of the service management and governance functions in OpenIoT. The un-deployment process is therefore addressed in later paragraphs as well.

5.3 Sensing-as-a-Service Infrastructure Anatomy

5.3.1 Lifecycle of a Sensing-as-a-Service Instance

As part of the OpenIoT system, the management and requests operations for dynamically creating and deploying IoT services (i.e. Sensing-as-a-Service and IoT-Analytics-as-a-Service services) perform the following main tasks:

- **Formulation of the request**: As part of this task the request is formed on the basis of the specification criteria for particular sensor selection, as well as of the processing of the resulting collected data.
- **Parsing and validation of the request**: This task processes the request and ensures its validity. The validation of the requests ensures that they refer to existing sensors and ICOs or that the criteria set lead to the selection of a group of sensors and ICOs.

- **Discovery of resources**: In the scope of this task the criteria for selecting sensors are applied against the OpenIoT directory services i.e. the sensor directory is used to select a set of sensors that fulfill the relevant criteria and when need it update the OpenIoT directory sensor services.
- **Instantiation of a new OpenIoT service**: With the selected sensors/ICOs at hand, a new OpenIoT service instance is created as a cloud service. This results in the establishment of the service that is associated with the Sensing-as-a-Service request.
- **Population of information and structures associated with utility metering and resource management**: Along with the creation of the OpenIoT service, the appropriate resources are reserved. This is denoted in the various structures that comprise information about the resources of the OpenIoT system. Furthermore, structures/records for the utility metrics are used.
- **Deployment/Delivery of the service**: As part of this task, the OpenIoT service is deployed and becomes available on the OpenIoT system. Consequently, it becomes ready to be invoked by end-users.

Figure 5.3 illustrates the main system actions entailed in the course/process of deploying an OpenIoT service (i.e. service request, sensor(s) selection, scheduling and resources reservation and ultimately service deployment). Following the successful deployment of an OpenIoT service, end-users can invoke and use it. As part of the service lifecycle, it is also likely that the service will be uninstalled and deactivated from the system, in which case all resources associated with the service will be released.

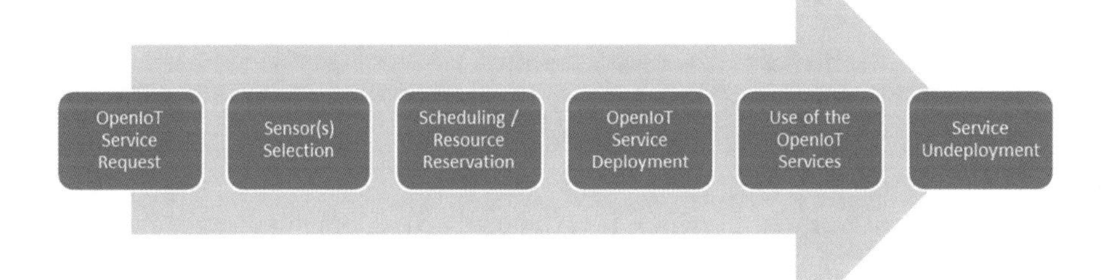

Figure 5.3 IoT data analysis services request lifecycle.

5.3.2 Interactions between OpenIoT Modules

OpenIoT is a sensor cloud environment. Along with the data of the various sensors and ICO streams, this cloud stores a wide range of meta-data enabling the deployment, delivery and optimization of IoT services within the sensor cloud. This meta-data is updated during the operation of the sensor cloud system, as new services are requested and deployed, while others go out of scope. The sensor cloud system will be responsible to frequently check if data are required from the system's deployed services from the provided mechanisms. Furthermore, this meta-data will regulate the interactions between the various components of the OpenIoT architecture. Figure 5.4 [10] illustrates the various modules of the OpenIoT architecture, along with their interactions (indicated based on uni-directional and bi-direction arrows). Furthermore, the figure illustrates the various entities/classes, whose values/data are used in the scope of the interactions of the modules. In particular, given the entities illustrated in Figure 5.4, each of the OpenIoT modules interacts with the others as follows:

- **Request Definition**: The request definition module is the user interface that enables the user to formulate the requests in the OpenIoT system. This module interacts directly with the Scheduler's API which is described in detail in following sections.
- **(Global) Scheduler**: The Scheduler formulates the request based on the user inputs (request definition). It interacts with the rest of the OpenIoT platform through the Cloud Database (DB). In particular, the Scheduler performs the following functions:
    - Retrieving the available sensors from the GSN nodes through the "availableSensors" entity,
    - Informing the GSN nodes abut which of their virtual sensors are used by the service being scheduled. Relevant information is includes in the "sensorServiceRelation" entity,
    - Informing the Service Delivery & Utility Manager (SD&UM) about what services to deliver based on the "serviceDeliveryDescription" entity,
    - Notifying the user, via itself and the SD&UM module, about the status of a defined service through the "serviceStatus" entity, and
    - Implementing access control mechanisms with the help of the "user" entity.

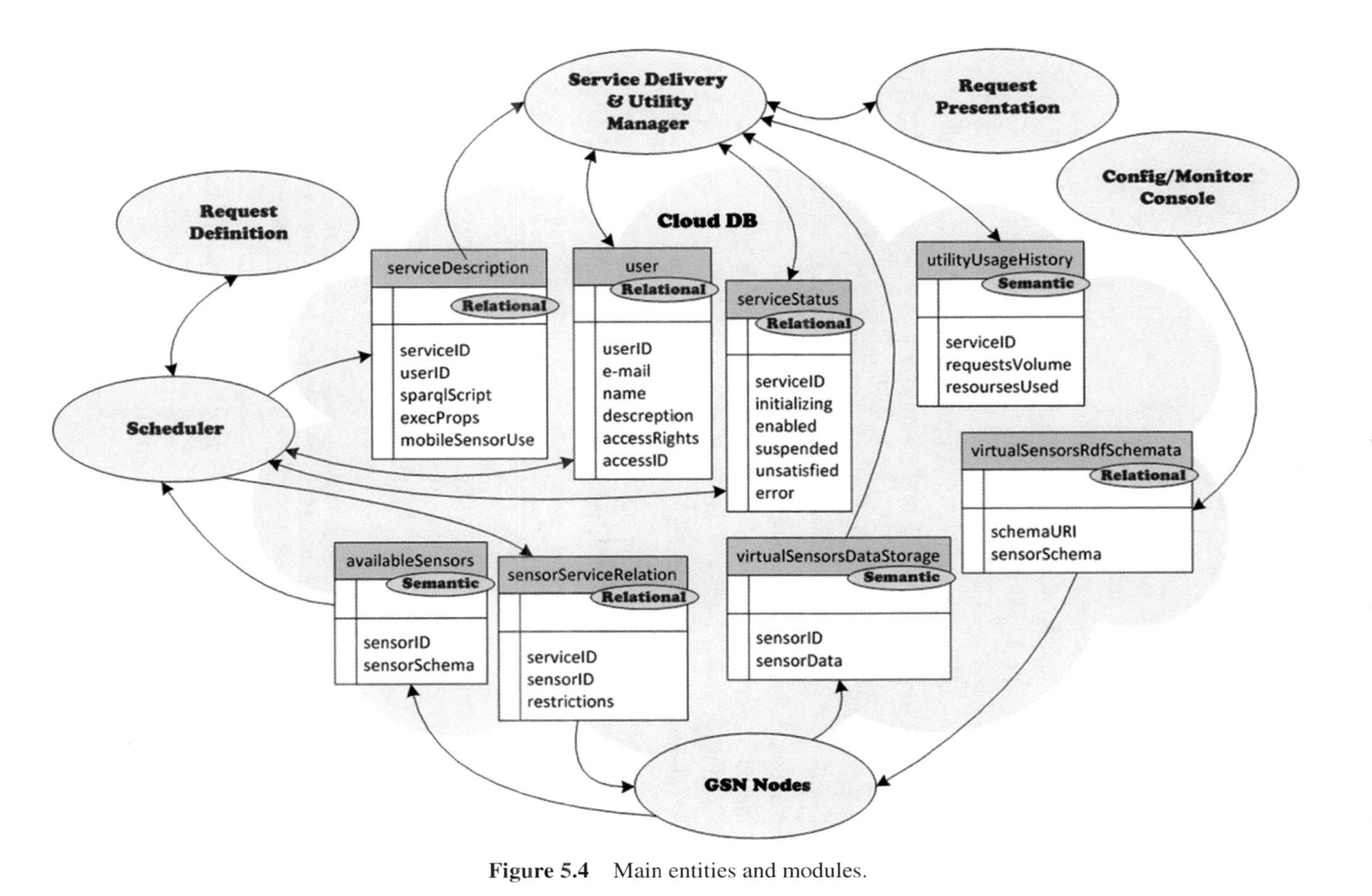

Figure 5.4 Main entities and modules.

- **Service Delivery & Utility Manager**: The SD&UM module provides results to the request presentation module by retrieving SPARQL scripts [7], that the Scheduler has provided to the "serviceDescription" entity. Furthermore, this module retrieves data from the GSN nodes by executing the retrieved scripts to the "virtualSensorsDataStorage". Moreover it is able to store resource-usage history for accounting, metering and billing purposes.
- **Request Presentation**: the Request Presentation module is the User Interface that enables the user to retrieve data from the Cloud Database (DB). The Request Definition has described the request and the data is delivered using the SD&UM API described below.
- **Configuration Console**: The Configuration/Monitoring console is the system administrators' tool, which enables administrators to deploy, configure and manage the OpenIoT platform. It interacts directly with several other modules (Scheduler, SD&UM and GSN nodes) for monitoring purposes. Finally, it is also capable to set up RDF schemata for new virtual sensors. The schemata are stored within the "virtualSensorsRdfSchemata" entity and enable GSN nodes to access this information during their configuration.
- **GSN Nodes**: The GSN nodes (or virtual sensors) are:
  - Providing the available sensors to the Scheduler module through the "availableSensors" entity,
  - Informed about the sensors in use from the Scheduler based on the "sensorServiceRelation" entity,
  - Retrieving new virtual Sensors RDF schemata from the Config/ Monitor Console through the "virtualSensorsRdfSchemata" entity, and
  - Providing sensor data to the SD&UM through the "virtualSensorsDataStorage" entity.

As part of these interactions the above modules create and consume data associated with the entities listed in the following table [10]. Note that the table differentiates between semantic and non-semantic data entities. Semantic data entities are implemented on the basis of ontologies (i.e. RDF), while non-semantic data structures are represented on the basis of relational database tables. Note that all the structures that hold sensor information follow semantic descriptions, given that all sensor descriptions in OpenIoT will be semantically annotated and represented.

Data Entity	Type	Description
serviceDescription	Relational (SQL) or RDF	Holds the description and properties of all the services that are executed through the OpenIoT system
availableSensors	Semantic (RDF)	Constitutes the directory database of the OpenIoT sensor cloud system.
serviceStatus	Relational (SQL)	Maintains a list with the status of the services, in order to provide relevant feedback to end-users
sensorServiceRelation	Relational (SQL)	Maintains the (many-to-many) associations of the services to the various sensors and ICOs available in the system (i.e. information about which sensors are used in the scope of a given services).
virtualSensorsDataStorage	Semantic (RDF)	Maintains the data of the various data streams i.e. data corresponding to the data streams of the sensors and ICOs that provide services to OpenIoT users
virtualSensorsRdfSchemata	Semantic (RDF)	Holds the structure of specific sensors/ICO types to allow for the management and instantiation of the sensors.
utilityUsageHistory	Semantic (RDF)	Used to records utility/usage related parameters, in order to boost accounting, billing and (utility based) resource optimization
user	Relational (SQL) or RDF	Used to store the available users and their access rights to implement access control mechanisms.

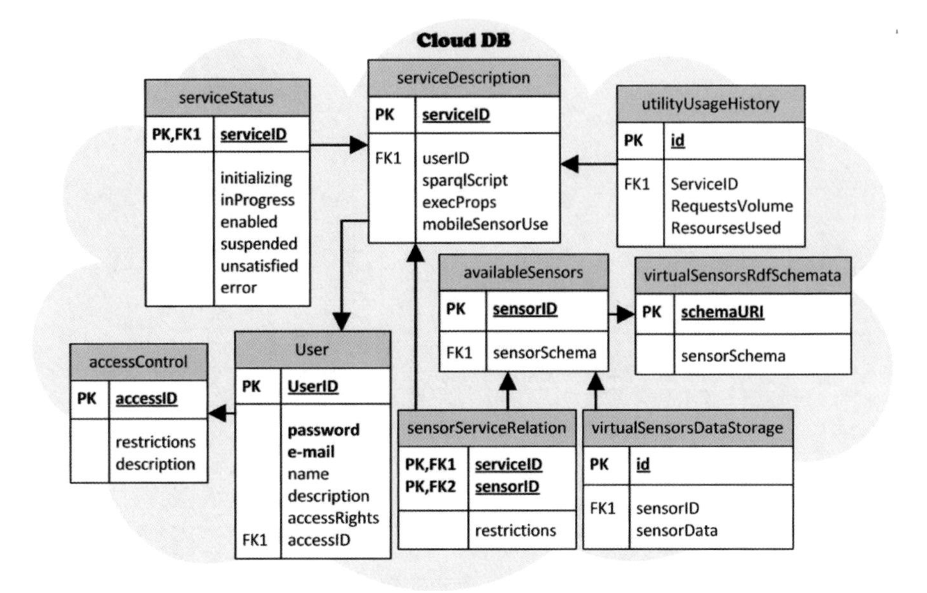

Figure 5.5 Relationships between the main OpenIoT data entities.

The relationship between the main OpenIoT data entities is depicted in Figure 5.5 [10].

5.4 Scheduling, Metering and Service Delivery

The modules that are responsible for the services formulation within the OpenIoT platform are the "Scheduler" and the "Service Delivery & Utility Manager". Following paragraphs provide a detailed description of these modules, including the functionalities that they offer to end-users. Note that the term end-user can either denote the final user of the IoT services or the solution provider exploiting the OpenIoT capabilities in order to integrate and deploy a Sensing-as-a-Service solution.

5.4.1 Scheduler

The Scheduler is the main and first entry point for service requests submitted to the OpenIoT cloud environment. This component receives the service requests from the service definition components as part of the process of creating a new cloud service based on the Sensing-as-a-Service paradigm. It parses each service request and accordingly performs two main functions towards the delivery of the service, the sensor/ICO selection and the scheduling/resource reservations.

The API of the scheduler supports the lifecycle of the OpenIoT service, which has been presented in earlier paragraph. In particular, it provides the means for:

- Constructing an OpenIoT service on the basis of existing sensors and ICOs.
- Registering an OpenIoT service within the OpenIoT sensor cloud. In this case the OpenIoT system assigns a service identifier (serviceID) to the service, which uniquely identifies the service within the OpenIoT service delivery system.
- Unregistering a (previous registered) OpenIoT service. This is a counterpart function to the one registering the service. The unregistration/deregistration function moves the service out of the scope of the OpenIoT system.
- Enabling an already registered service, thereby commencing its operation within the OpenIoT sensor cloud.
- Disabling an OpenIoT service, thereby leading to its deactivation within the sensor cloud. Disabling a service does not however imply that the service goes out of the scope of the sensor cloud i.e. it still remains available for activation.
- Querying the status of a given service, as a means of accessing the state of the service within the sensor cloud.

The above functions change the state of the OpenIoT services according to rules and dependencies specified within the various states. For example, only registered services can be enabled, and only enabled services can be disabled. At the same time, only registered services can be unregistered.

Figure 5.6 [8] illustrates the lifecycle of the IoT services within the OpenIoT system. The transitions between the different states occur on the basis of invocations to the Scheduler API.

On the basis of the Scheduler API, the following functionalities are supported:

- **Resource Discovery**: This service will discover virtual sensor availability based on the "availableSensors" entity. It will provide the resources that match the requirements for a given service request.
- **Service User Management**: This Scheduler service will enable the management of the lifecycle of an OpenIoT service. This lifecycle management is performed based on the following Scheduler comments:

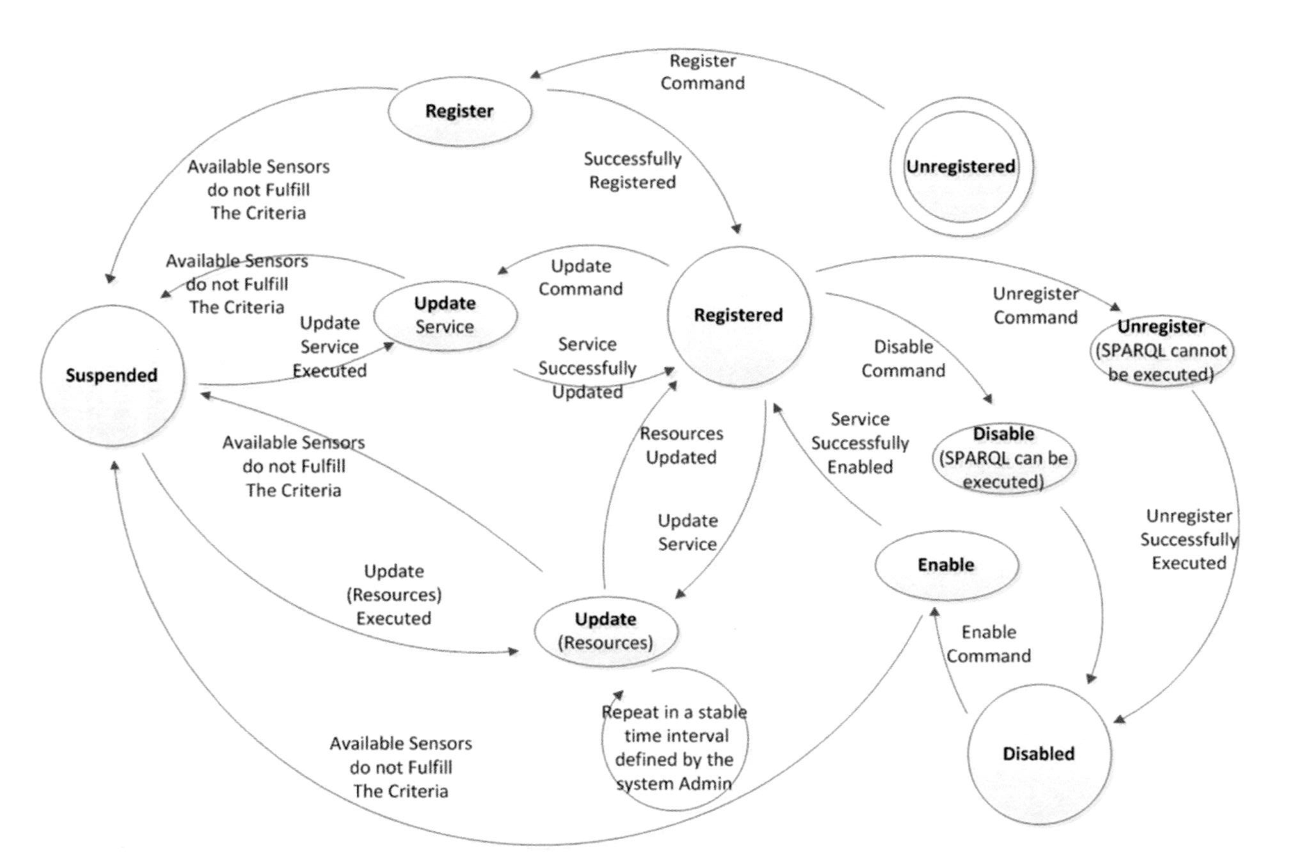

Figure 5.6 State diagram of the OpenIoT services lifecycle within the scheduler module.

- *Register*: The “Register” service is responsible to identify all the required resources from the request and update the “sensorService-Relation” entity at the cloud database. The “Register” service shall formulate a SPARQL script, based on the user request, and shall store it to the “ServiceDescription” entity along with a Service ID and user’s specific execution properties (the execution properties could include execution intervals, life of the service, etch). A new service instance shall get recorded at the “serviceStatus” entity in the cloud. Note that: (a) In case the request is satisfied the unsatisfied Boolean of the “serviceStatus” entity is set to false, whereas (b) If the request is unsatisfied the unsatisfied Boolean of the “serviceStatus” entity is set to true. Optionally more detailed information regarding the problem could be stored.
- *Unregister*: In the scope of the unregister functionality the user will have the ability to unregister a registered service. When a service gets unregistered the allocated resources shall get released. Therefore, the service-virtual sensor relation at the “SensorService-Relation” entity in the cloud shall get deleted. Furthermore, the service gets deactivated (set enabled as false) at the “serviceStatus” entity in the cloud.
- *Suspend*: As part of suspend functionality, the service shall get updated (set suspended as true) at the “serviceStatus” entity in the cloud.
- *Enable from Suspension*: As part of the suspension functionality the service is defined as enabled (enabled is true) at the “serviceStatus” entity.
- *Enable*: The enable functionality gives to the user will be given the ability to enable an unregistered service. When a service gets enabled the user request gets initialized and the related virtual sensors are identified and stored to the “SensorServiceRelation” entity. The service is set as enabled at the “serviceStatus” entity.
- *Update*: The update services permits changes to service. When a registered service gets updated the “Update” identifies all the required resources from the updated request and updates the “Sensor-ServiceRelation” entity at the cloud database. The “Update” service shall formulate a SPARQL script, based on the updated user request, and shall update it to the existing one along with the updated user’s specific execution properties (the execution

properties could include execution intervals, life of the service, etch) at the "ServiceDescription" entity. The service status shall get updated as enabled at the "serviceStatus" entity in the cloud. Note that: (a) In case the request is satisfied the unsatisfied Boolean of the "serviceStatus" entity is set to false, (b) In case the request is unsatisfied the unsatisfied Boolean of the "serviceStatus" entity is set to true. Optionally more detailed information regarding the problem could be stored.

- **Registered Service Status**: This functionality enables the user to retrieve the status of a specific service by providing the ServiceID. The Registered Service Status service shall check the "serviceStatus" entity and send all the available information back to the user.
- **Service Update Resources**: Based on a service provider (i.e. administrator controlled) specified time interval this service/functionality shall check the enabled services from the "serviceDescription" entity and as a first step identify the ones that are using mobile sensors. As a second step it shall check if the mobile sensors fulfil the User's request (e.g. in respect of a specific location). Note that: (a) in case the sensor fulfills the user's request no further action is taken and (b) in case the sensor does not fulfil the user's request this sensor is unrelated/removed from the specific service at the "sensorServiceRelation" entity and (c) as a third step a new sensor is searched that fulfils the user's request (e.g. in respect of a specific location), (d) in case a new sensor is found it gets recorded at the "ServiceDescription" entity and the "serviceDescription" entity gets updated, (e) in case is no sensor available that fulfils the specific request the unsatisfied field shall get updated with "true" at the "serviceStatus" entity in the cloud.
- **Get Service**: This service is used to get the description of a registered service. Accessing the "serviceDescription" entity retrieves this information.
- **Get the Available Services**: This service provides the ability to a user to collect a list of registered services related with a specific user. These service IDs are available from the "serviceDescription" entity.
- **Get User**: This service is used by the OpenIoT platform's access controls mechanisms so as to retrieve user's information, access rights and restrictions to implement data filtering and access rights.

Note that for the user to be able to invoke the "Resource Discovery", "Service User Management", "Registered Service Status", "Service Update

Resources", "Get Service", "Get User" and "Get the available Services" services, the user must first get logged-in to the system by authenticating with his/her ID. Moreover, the results provided to the user are prior filtered based on his/her account restrictions and the resources that are accessible based on his/her profile. The "user" and the "accessControl" entities provide the account restrictions data.

In line with the Scheduler functionalities presented above, Figure 5.7 [8] illustrates the main workflow associated with the service registration process. In the scope of this process the Scheduler attempts to discover the resources (sensors, ICO) that will be used for the service delivery. In case no sensors/ICOs can fulfill the request, the service is suspended. In case a set of proper sensors/IOCs is defined the relevant data entities are updated (e.g., relationship of sensors to services) and a SPARQL script associated with the service is formulated and stored for later use. Following the successful

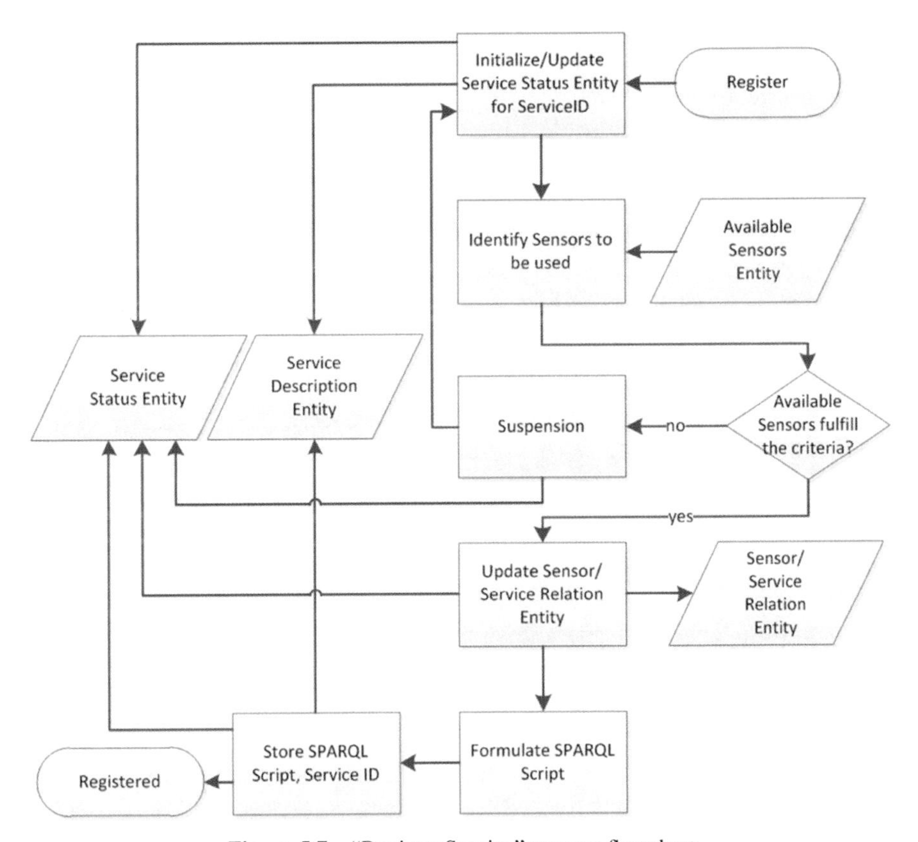

Figure 5.7 "Register Service" process flowchart.

conclusion of this process, the servicer enters the «Registered» state and is available for invocation.

Likewise Figure 5.8 [8] illustrates the process of updating the resources associated with a given service. As already outlined, such an update process is particularly important when it comes to dealing with IoT services that entail mobile sensors and ICOs i.e. sensors and ICOs whose location is likely to change within very short timescales (such as mobile phones and UAV (Unmanned Aerial Vehicles)). In such cases the update resources process could regularly check the availability of mobile sensors and their suitability for the registered service whose resources are updated. The workflow in Figure 5.7 assumes that the list of mobile sensors is known to the service (i.e. the sensors' semantic annotations indicate whether a sensor is mobile or not).

Even though the process/functionality of updating resources is associated with the need to identify the availability and suitability of mobile sensors, in principle the update process could be used to update the whole list of resources that contribute to the given service. Such functionality could help OpenIoT in dealing with the volatility of IoT environments, where sensors and ICOs may dynamically join or leave. In the scope of an IoT application, one cannot rule out the possibility of the emergence of new sensors that can be associated with an already established service.

Finally, Figure 5.9 [8] illustrates the process of unregistering a service, in which case the resource associated with the service is released. The data structures of the OpenIoT service infrastructures are also modified to reflect the fact that the specified service no longer using its resources. As already explained, this update is important for the later implementation of the OpenIoT self-management and optimization functionalities.

5.4.2 Service Delivery & Utility Manager

The Service Delivery & Utility Manager has (as its name indicates) a dual functionality. On the one hand (as a service manager) it is the module enabling data retrieval from the selected sensors comprising the OpenIoT service. On the other hand, the utility manager maintains and retrieves information structures regarding service usage and supports metering, charging and resource management processes. The following paragraphs elaborate on the main functionalities/services of the Service Delivery & Utility Manager.

The API of the Service Delivery & Utility Manager (SD&UM) serves as the point where the OpenIoT platform provides its outcome. In particular, the module provides the means for:

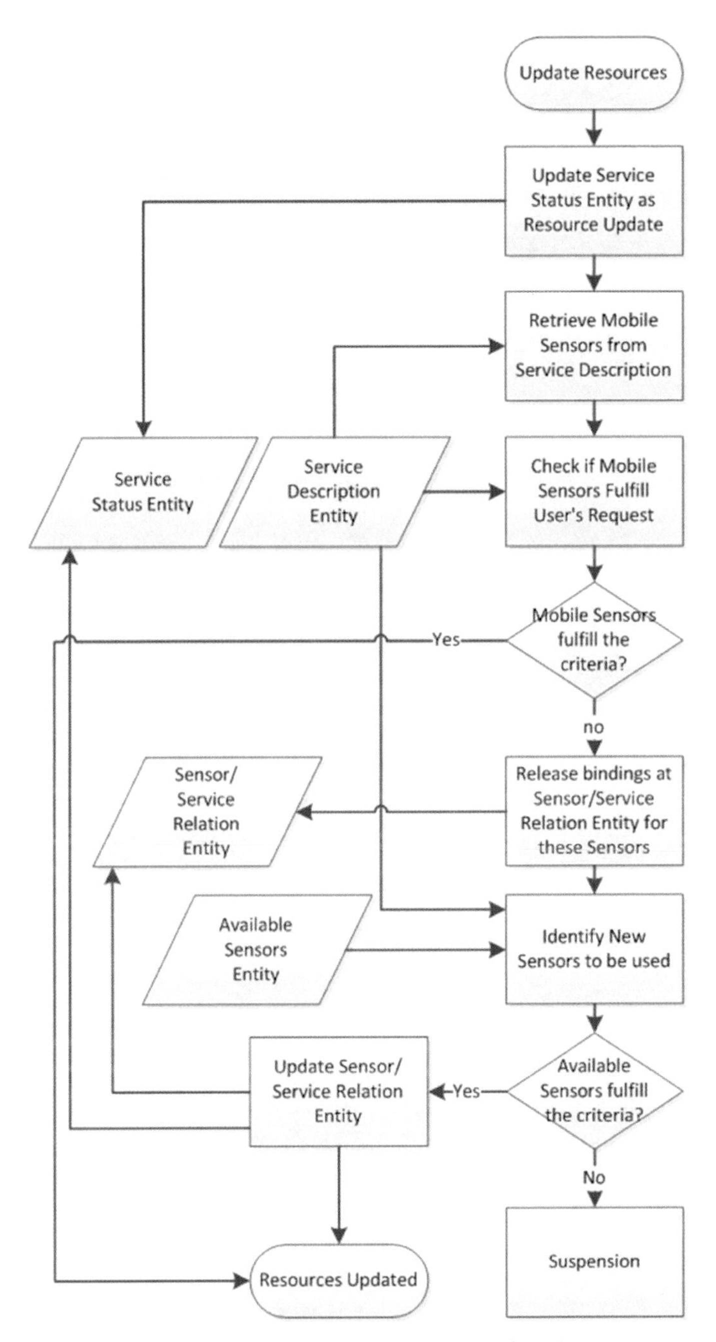

Figure 5.8 "Update Resources" service flowchart.

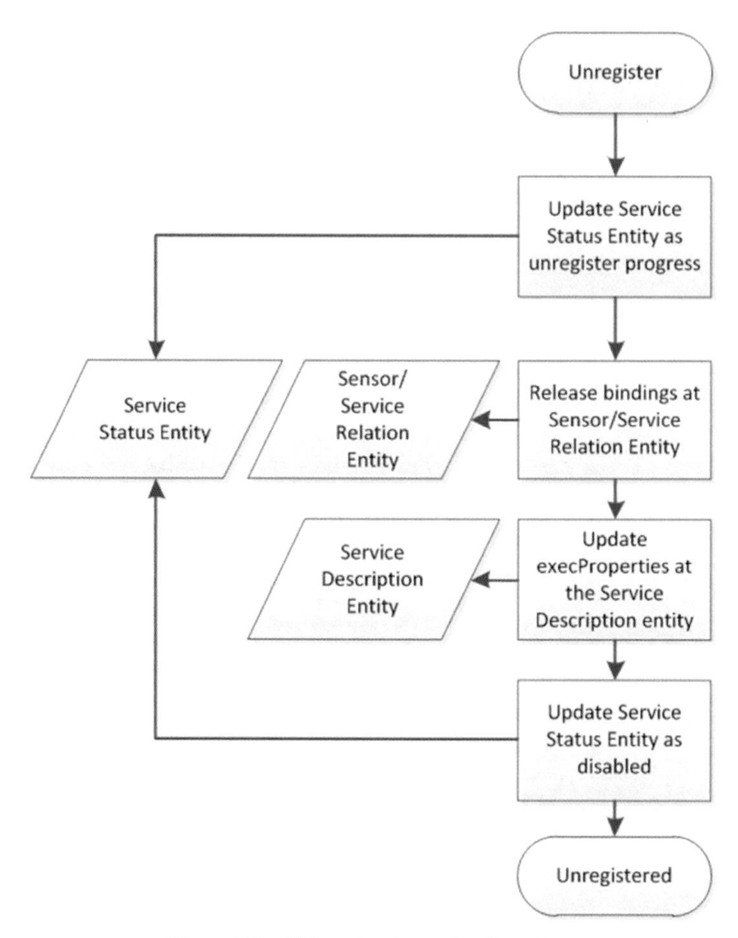

Figure 5.9 "Unregister" service flowchart.

- Executing and delivering the requested services.
- Accessing and processing data streams from the cloud.
- Taking into account processing instructions specified during the request formulation.
- Keeping track of utility parameters associated with the service, for example: the time the service is used, the volume of data transmitted, as well as the number and type of sensors used.
- Managing and maintaining utility data records.

On the basis of the Service Delivery & Utility Manager API, the following functionalities are supported:

- **Subscribe for a report**: This service enables the user to invoke an already defined service from the "ServiceDescription" entity. By providing an application's destination address (URI) this service will collect the results from the predefined query (sparqlScript), which is stored at the "ServiceDescription" entity, and deliver it to the application via the Callback Service.
- **Callback Service**: this service is instantiated by the "Subscribe for a report" service and invoked based on the schedule defined by the user at the service registration time. If the query is executed normally, the callback service invokes the callback results service.
- **Callback results**: By invoking the callback results the SD&UM will attempt to deliver results to the subscriber application.
- **Unsubscribe for a report**: This service is invoked by the user and deactivates the "Subscribe for a report" one. The previously registered subscription removal is identified by the user by providing a unique subscription ID.
- **Poll for a report**: This service enables the user to invoke an already defined service from the "serviceDescription" entity. The difference with the "subscribe for a report" service is that it enables the user to execute the predefined query with modified parameters (i.e. give me the results of the last 30 min) and that this call will produce a single Result Set (it will be executed only once and then it will be dropped). In case the query executes normally, the "Poll for a report" service invokes the callback results service.
- **Get the utility usage of a user**: This service enables the user to retrieve the utility usage involved for a specific user. By providing the user's ID the "Get the utility usage of a user" service retrieves the related services with the specific user from the "serviceDescription" entity. It then collects the usage history from the "utilityUsageHistory" entity and by using special utility usage algorithms and in relation with the policies applied for the provided services, it returns the overall usage/cost of the platform for the selected user.
- **Get the utility usage of a registered service**: This service enables the user to retrieve the utility usage related with a specific registered service. By providing the "serviceID" it collects the usage history from the "utilityUsageHistory" entity and by employing special utility usage algorithms combined with the charging policies specified for the provided services it returns the usage/cost of the platform for the selected service.

- **Record utility usage of a service**: This service is invoked from the "Poll for a report" and the "Callback service" services. On its invocation the volume of the requested data and the type of resources used, are stored to the "utilityUsageHistory" entity for later use from the "Get the utility usage of a registered service" and the "Get the utility usage of a user" services.
- **Get service status**: This service enables the user to retrieve the status of a specific service by providing the service ID. The registered service status service shall check the "serviceStatus" entity and send to the user all the available information.
- **Get service**: This service is used to get the description of a registered service. This information is retrieved by accessing the "serviceDescription" entity.
- **Get the available services**: This service provides the ability to a user to collect a list of registered services related with a specific user. These service IDs are available from the "serviceDescription" entity.
- **Get User**: This service is used by the OpenIoT platform's access controls mechanisms so as to retrieve a user's information, access rights and restrictions in order to implement data filtering and access rights.

Note that to be able to invoke the "Subscribe for a report", "Unsubscribe for a report", "Poll for a report", "Get the Utility Usage of a User", "Get Service", "Get User" and "Get the available Services" services the user must first get logged-in to the system by authenticating with his/her ID. Moreover the results provided to the user are prior filtered based on his/her account restrictions and the resources which are accessible based on his/her profile. The account restrictions data are provided by the "user" and the "accessControl" entities.

5.5 Sensing-as-a-Service Example

Following paragraphs illustrate the process of establishing a fully deployable service (from data Capturing to Visualization) using the OpenIoT reference framework and its Sensing-as-a-Service capabilities.

5.5.1 Data Capturing and Flow Description

In this example, weather sensors are deployed in the central area of Brussels producing data (wind chill temperature, atmospheric pressure, air temperature, atmosphere humidity and wind speed).

The data are captured using the GSN middleware[1] through a special wrapper (i.e. residing at the physical plane of the architecture depicted in Figure 5.1) which collects the Weather Station's data every 4 hours. This is where the first level of data filtering occurs, whereas the weather station produces data in a higher rate, in this scenario we are interested in a four hour sampling rate. The captured data are following a sensor type created for this occasion (named after "Weather"). The "Weather" sensor type is used to semantically annotate the captured data at the GSN level. One GSN instance is running for every weather station so after X-GSN announces the existence of each sensor (bound with a specific sensor id) it starts to push the captured data to Linked Sensor Middleware (LSM) components, which comprises an RDF Store and is deployed in a private cloud environment (the virtualized plane of the architecture Figure 5.1).

Then it is time to set up the service by using the Request Definition (the utility/application plane of Figure 5.1) tool with the help of which we will discover these sensors (by using the Scheduler), describe the request and send it to the Scheduler (the virtualized plane of Figure 5.1) to handle it. The Scheduler decomposes the request and registers it to LSM. The information that should be accessed and processes in this scenario is the wind chill temperature versus the actual air temperature in the area of Brussels for the dates between 01/07/2014 and 01/28/2014.

The SD&UM (the virtualized plane in Figure 5.1) retrieves on demand the formulated request executes the involved queries and feeds the Request Presentation (i.e. the utility/application plane of the OpenIoT architecture) with presentation data. The last step would be for the Request Presentation to presents the received data in the predefined widgets.

The presented high level description of the data flow at the virtualized and utility/application planes is in following paragraphs built and presented as an OpenIoT Sensing-as-a-Service application.

5.5.2 Semantic Annotation of Sensor Data

The association of metadata with a virtual sensor is performed through an appropriate metadata file. For example, a virtual sensor named Brussels_weather.xml will have an associated metadata file named Brussels_weather.

[1]Also called X-GSN (extended GSN) in the context of OpenIoT, where an enhanced version of the original GSN middleware that supports semantic annotation of virtual sensors has been deployed.

metadata. The metadata file contains information such as the location (in coordinates), as well as the fields exposed by the virtual sensor. This also includes the mapping between a sensor field (e.g. airtemperature) and the corresponding high-level concept of the ontology (e.g., http://openiot.eu/ontology/ns/AirTemperature).

```
sensorID="http://lsm.deri.ie/resource/61330620147099"
sensorName=979128
source="Brussels netatmo"
sensorType=weather
information=Weather sensors in Brussels
author=openiot
feature="http://lsm.deri.ie/OpenIoT/BrusselsFeature"
fields="pressure,airtemperature,humidity,visibility,win
dchill,windspeed"
field.airtemperature.propertyName="http://openiot.eu/on
tology/ns/AirTemperature"
field.airtemperature.unit=C
field.humidity.propertyName="
http://openiot.eu/ontology/ns/AtmosphereHumidity"
field.humidity.unit=Percent
field.visibility.propertyName="
http://openiot.eu/ontology/ns/AtmosphereVisibility"
field.visibility.unit=Percent
field.pressure.propertyName="
http://openiot.eu/ontology/ns/AtmosphericPressure"
field.pressure.unit=mb
field.windchill.propertyName="
http://openiot.eu/ontology/ns/WindChill"
field.windchill.unit=C
field.windspeed.propertyName="
http://openiot.eu/ontology/ns/WindSpeed"
field.windspeed.unit=Km/h
latitude=51.33332825
longitude=3.200000048
```

5.5.3 Registering Sensors to LSM

Sensors can be registered to the LSM middleware (and its cloud datastore) by executing an appropriate script (i.e. lsm-register.sh (on Linux/Mac) or lsm-register.bat (on Windows)). This script takes as argument the metadata file name. After this, the corresponding metadata in RDF will have been stored in LSM. An example is illustrated in the following table:

```
./lsm-register.sh virtual-
sensors/brussles_weather.metadata
lsm-register.bat virtual-
sensors\brussels_weather.metadata
```

5.5.4 Pushing Data to LSM

In order to push data to LSM, the LSMExporter processing class is internally used by GSN/X-GSN. This is specified in the virtual sensor configuration file:

```
<processing-class>
      <class-name>org.openiot.gsn.vsensor.LSMExporter
</class-name>
      <init-params>
          <param name="allow-nulls">false</param>
          <param name="publish-to-lsm">true</param>
      </init-params>
      <output-structure>
          <field name="airtemperature" type="double" />
          <field name="humidity" type="double" />
          <field name="pressure" type="double" />
          <field name="windspeed" type="double" />
          <field name="windchill" type="double" />
          <field name="visibility" type="double" />
       </output-structure>
</processing-class>
```

Then, when X-GSN starts, it begins to acquire the data through the wrapper and automatically generating the RDF data for each observation, storing it in LSM.

Each observation will be assigned a unique URI, e.g.

<http://lsm.deri.ie/resource/29925179667811>

Then, you can query the Virtuoso server, to see the updated data, with the SPARQL query shown in the following table:

```
select * where {
  <http://lsm.deri.ie/resource/29925179667811> ?p ?o
}
```

and get the results shown in the following table:

http://www.w3.org/1999/02/22-rdf-syntax-ns#type	http://purl.oclc.org/NET/ssnx/ssn#Observation
http://purl.oclc.org/NET/ssnx/ssn#observedBy	http://lsm.deri.ie/resource/29855158254802
http://purl.oclc.org/NET/ssnx/ssn#observationResultTime	2013-05-15T11:45:00Z
http://purl.oclc.org/NET/ssnx/ssn#featureOfInterest	http://lsm.deri.ie/resource/3797289123726234

Once the data is in LSM, it can be accessed by the other OpenIoT components.

5.5.5 Service Definition and Deployment Using OpenIoT Tools

The first step, towards building a request for Sensing-as-a-Service, would be to log in to the Request Definition by using our credentials (Figure 5.10).

By logging in our profile is loaded and all our previously defined services are available to view or edit (Figure 5.11). A new Application can be created through the "File" menu (Figure 5.12).

As a first step, the available sensors should be discovered, using the magnifying glass at the data sources toolbox. In the map that appears we look up for the Brussels area and we add a pinpoint to the map. Then we set the radius of interest and we hit the "Find sensors" button (Figure 5.13 [9]).

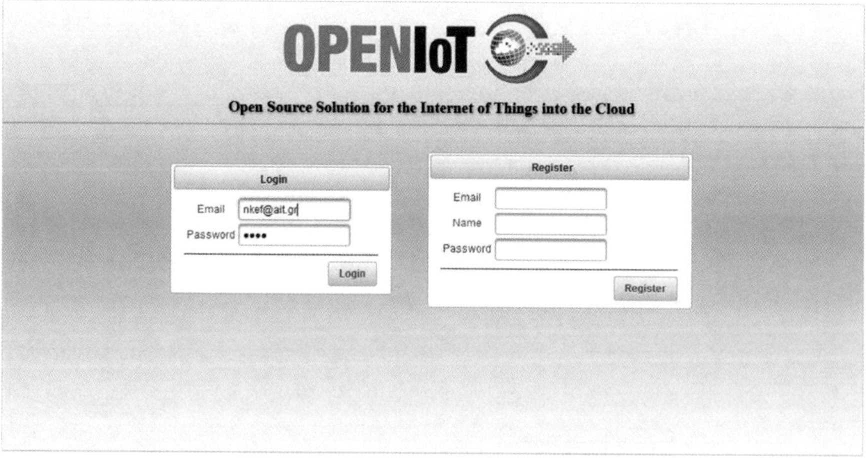

Figure 5.10 Request definition log in.

Figure 5.11 Request definition loaded profile.

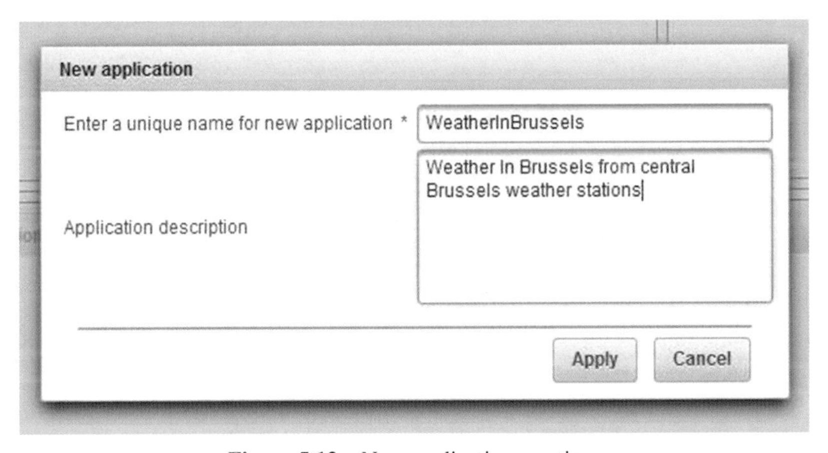

Figure 5.12 New application creation.

This request is send to the Scheduler that in its turn queries LSM for available sensors in this area. The reported, from LSM, sensor types are sent to the Scheduler that in its turn sends to the Request Definition so as to fill the available "Data sources" toolbox (Figure 5.14). As we can see two sensor

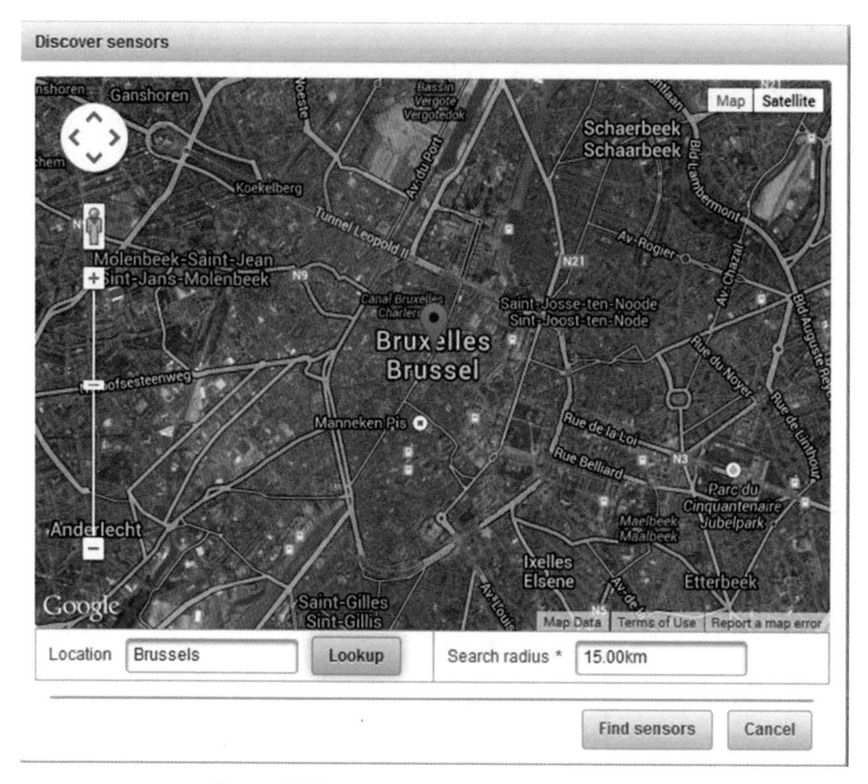

Figure 5.13 Sensor discovery in Brussels area.

types are deployed in that area (weather sensors and Integra Traceability Kiosk sensors).[2] By dragging and dropping the blocks from our toolbox we start to build our request. We drag and drop the "weather" sensor type and as we can see all the sensor type observations (outputs) are available to interact with (wind chill temperature, atmospheric pressure, air temperature, atmosphere humidity and wind speed).

A "Selection filter" from the "Filters & Groupers" toolbox is required. The one side of it is connected with the node and the other one with a "Between" comparator that has already been dropped to the workspace from the "Comparators" toolbox. We set up the "Between" comparator between "01/07/2014" and "01/28/2014" (three weeks) which are the dates of interest

[2]ITK is a multi-sensor device for track & trace applications in manufacturing and used in the scope of other OpenIoT applications.

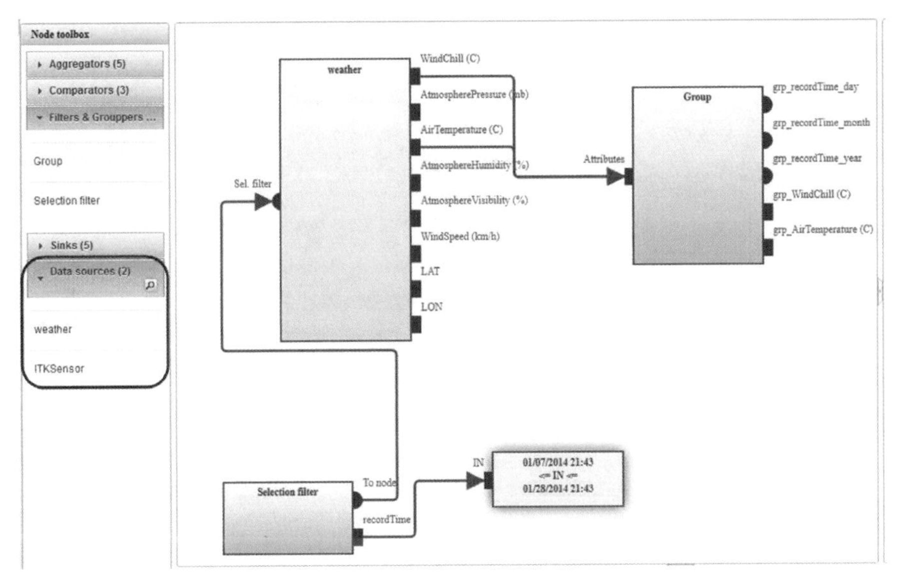

Figure 5.14 Comparator (between) properties.

to us to collect our data (Figure 5.14). The next step is to add a "Group" node from the "Filters & Groupers" toolbox which we are going to use so as to group the Wind Chill and Air Temperature by Year/Month/Day (see Figure 5.15 [9]) which is selected through the node's options. The Wind chill and Air temperature outputs of the "weather" node are connected to the "Group" node attributes and automatically. As shown in Figure 5.16, these outputs are generated also to the "Group" node.

Since we need the average values for every day, we drag and drop two "Average" nodes from the "Aggregators" toolbox to the workspace and we connect the Wind Chill and Air Temperature outputs to them respectively (see Figure 5.16). The next step required in order to visualise the output (i.e. two average values for every day) to a line chart, is to drag and drop a "Line Chart" from the "Sinks" toolbox. The X axis presents the time and the Y axis presents/compares the temperature values. At the line chart properties, two series count are presented (in order to visualize two inputs) and for the X axis we select date observation as type. Hence, all the day/month/year outputs of the "Group" node are connected to "x1" and "x2" inputs of the "Line Chart" node respectively and the Wind Chill and Air Temperature outputs to "y1" and "y2" inputs respectively (Figure 5.16 [9]).

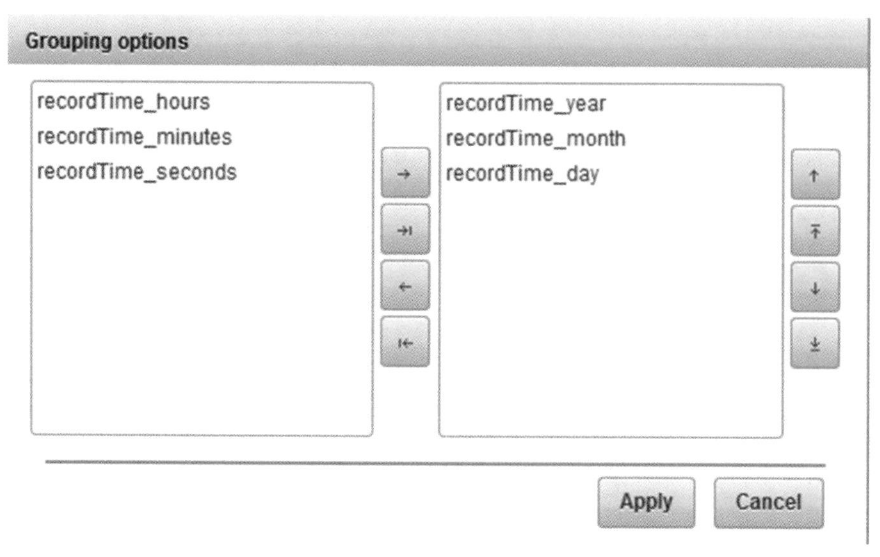

Figure 5.15 Grouping options.

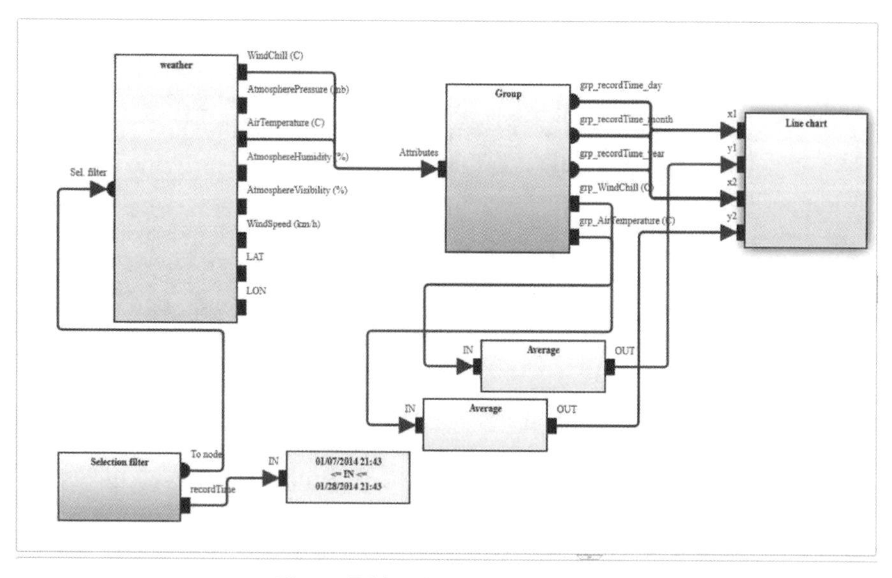

Figure 5.16 Line chart properties.

Following the visual definition of the service, the overall design can be validated using the "Validate design" option of the "Current application" menu (see Figure 5.17). This generates automatically the SPARQL scripts

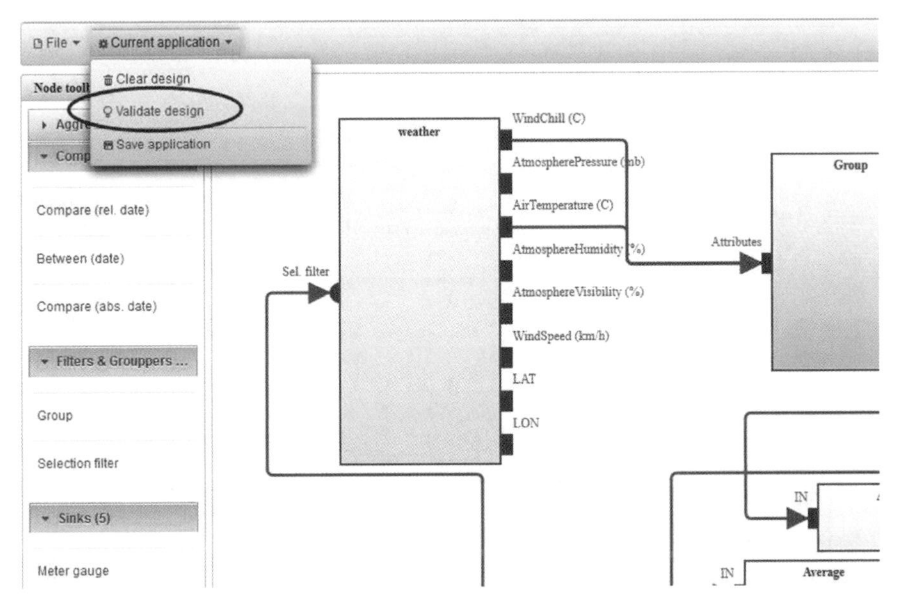

Figure 5.17 Validation of the service design.

that describe the graphical representation in our workspace. For every group of data that provides its output to a widget a different script is generated. In this specific example there is a need to visualize two different outputs (Wind Chill and Air Temperature) in one line chart and hence two scripts are generated (Figure 5.18 [9]).

For testing purposes these scripts could be taken and executed directly against the SPARQL interface of LSM (e.g., Figure 5.19).

The Request Definition UI can also be used to save (register) the newly described Sensing-as-a-Service application to the Scheduler (Figure 5.20).

5.5.6 Visualizing the Request

In order to visualize the captured data, one has to log-in to the Request Presentation UI. Following this log-in the user profile is loaded and the user is able to view all the services registered under his account. The registered services are fetched from the SD&UM, which also builds the appropriate scripts to query this information from LSM (Figure 5.21).

Then we choose the application of interest to us (i.e. "WeatherInBrussels") (Figure 5.22).

Accordingly, an empty widget associated with the selected application is presented. By using the "force dashboard refresh" option from the

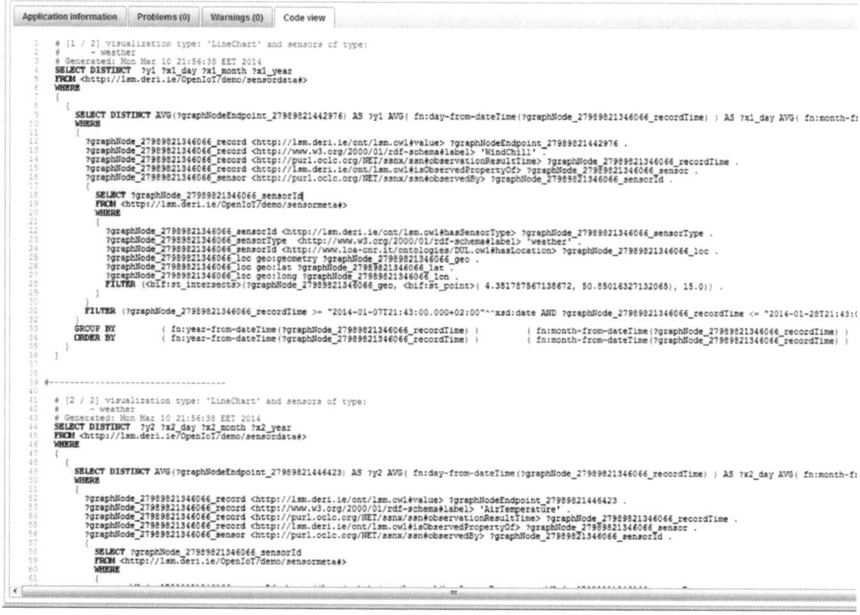

Figure 5.18 SPARQL script generation.

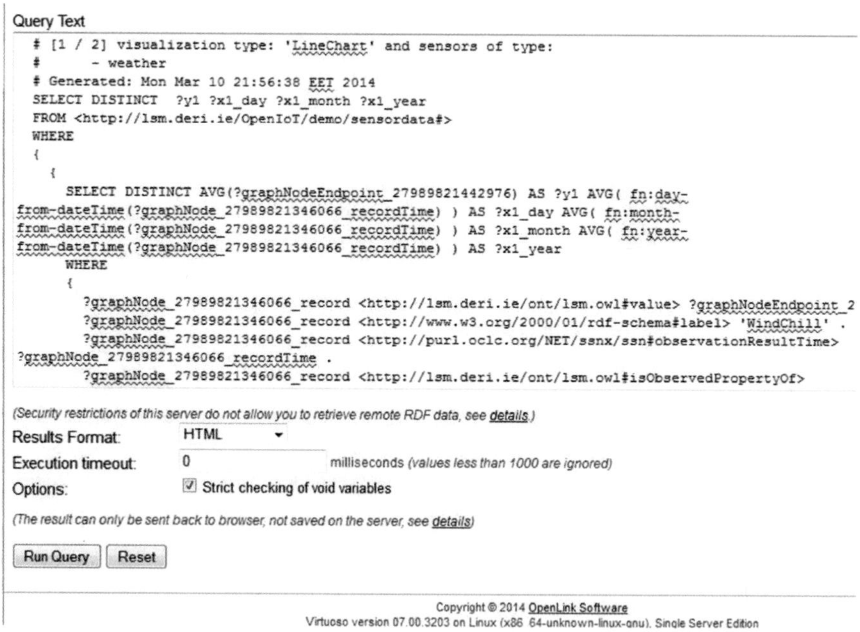

Query Text

```
# [1 / 2] visualization type: 'LineChart' and sensors of type:
#       - weather
# Generated: Mon Mar 10 21:56:38 EET 2014
SELECT DISTINCT  ?y1 ?x1_day ?x1_month ?x1_year
FROM <http://lsm.deri.ie/OpenIoT/demo/sensordata#>
WHERE
{
  {
     SELECT DISTINCT AVG(?graphNodeEndpoint_27989821442976) AS ?y1 AVG( fn:day-
from-dateTime(?graphNode_27989821346066_recordTime) ) AS ?x1_day AVG( fn:month-
from-dateTime(?graphNode_27989821346066_recordTime) ) AS ?x1_month AVG( fn:year-
from-dateTime(?graphNode_27989821346066_recordTime) ) AS ?x1_year
     WHERE
     {
        ?graphNode_27989821346066_record <http://lsm.deri.ie/ont/lsm.owl#value> ?graphNodeEndpoint_2
        ?graphNode_27989821346066_record <http://www.w3.org/2000/01/rdf-schema#label> 'WindChill' .
        ?graphNode_27989821346066_record <http://purl.oclc.org/NET/ssnx/ssn#observationResultTime>
?graphNode_27989821346066_recordTime .
        ?graphNode_27989821346066_record <http://lsm.deri.ie/ont/lsm.owl#isObservedPropertyOf>
```

(Security restrictions of this server do not allow you to retrieve remote RDF data, see details.)

Results Format: HTML

Execution timeout: 0 milliseconds (values less than 1000 are ignored)

Options: Strict checking of void variables

(The result can only be sent back to browser, not saved on the server, see details)

Run Query Reset

Copyright © 2014 OpenLink Software
Virtuoso version 07.00.3203 on Linux (x86_64-unknown-linux-gnu), Single Server Edition

Figure 5.19 LSM SPARQL endpoint (2 weeks wind chill in Brussels).

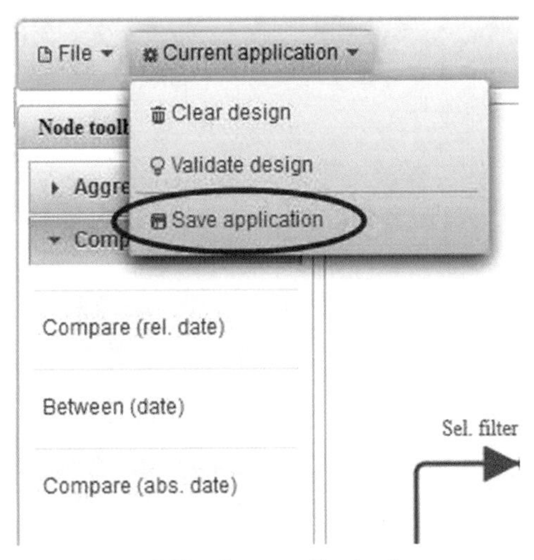

Figure 5.20 Save application button.

Figure 5.21 Request presentation loaded profile.

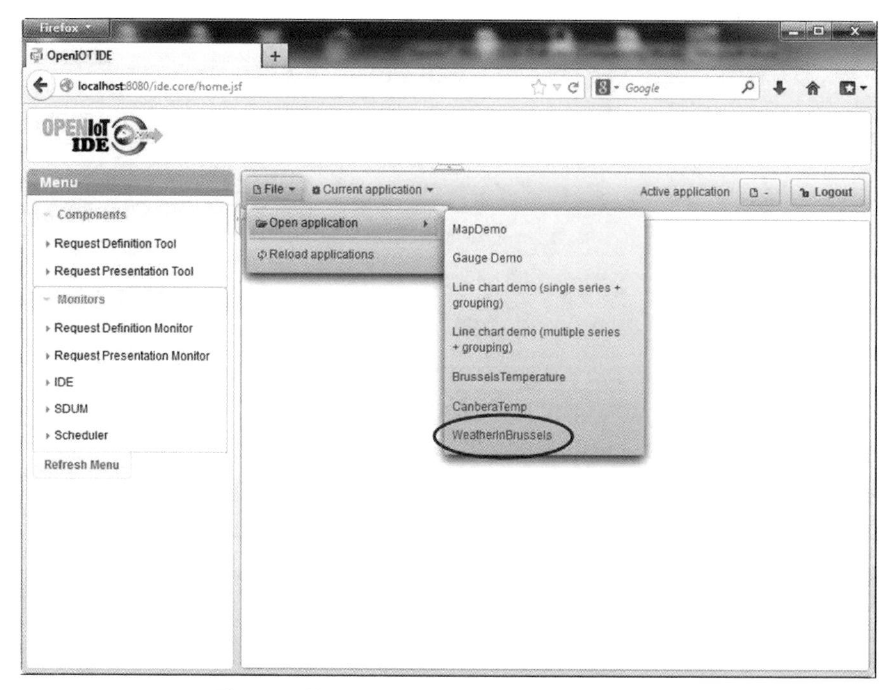

Figure 5.22 Load "WeatherInBrussels" scenario.

"Current application" menu, the Request Presentation exploits the "poll-ForReport (serviceID: String): SdumServiceResultSet" rest service of the SD&UM. This SD&UM service retrieves the previously registered application from the LSM module, retrieves the involved SPARQL scripts, executes them against the LSM SPARQL interface, analyses the results, builds a list of the results and how to present them to the widget and finally sends these data to the Request Presentation module where the result is visualized (Figure 5.23 [9]). The result is a filtered result set from the initially raw data stored to the database every 4 hours of the average Wind Chill temperature versus average Air temperature in Brussels for the specified time interval.

5.6 From Sensing-as-a-Service to IoT-Analytics-as-a-Service

Earlier paragraphs have illustrated the Sensing-as-a-Service paradigm, along with its practical implementation based on the OpenIoT open source project

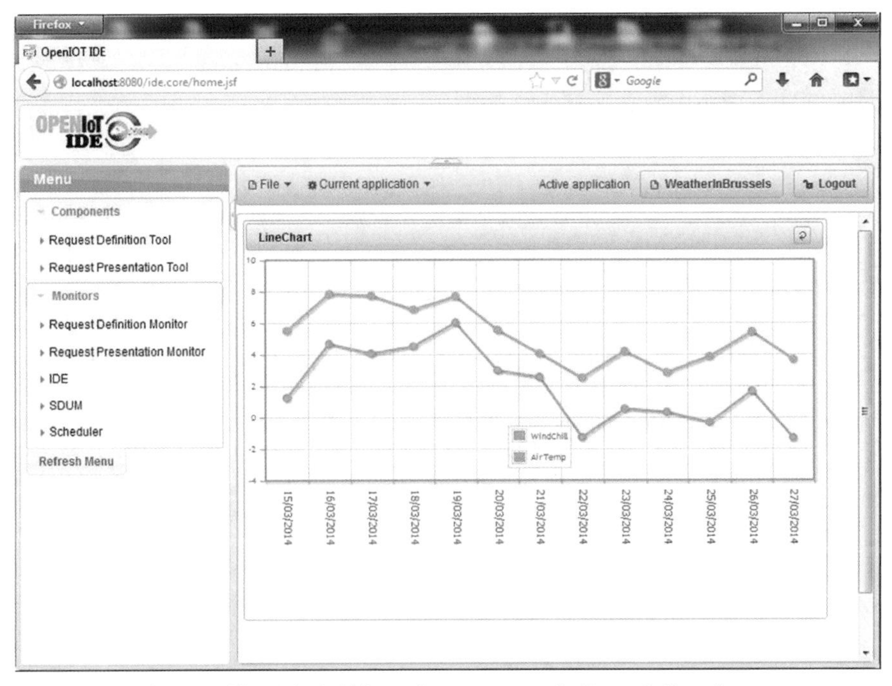

Figure 5.23 Wind chill vs. air temperature in Brussels line chart.

and the tools that it provides. The Sensing-as-a-Service paradigm as implemented by OpenIoT involves:

- Dynamic selection of (virtual) sensors from the set of sensors that are registered with the (RDF-based) directory services. This selection of sensors is empowered by the semantic unification of diverse data streams, which is supported by OpenIoT on the basis of the semantic annotation of virtual sensors and their observations.
- Definition of IoT data processing functions over the selected IoT data sources based on functionalities that can be expressed as SPARQL queries. Note that SPARQL does not enable the definition and execution of sophisticated data analytics functions. Rather, it is limited to supporting simple statistical processing functionalities such as the calculation of sums, averages and variances over observations provided by the selected virtual sensors and/or groups of virtual sensors.

Hence, the introduced Sensing-as-a-Service functionalities do not provide the means for non-trivial data analytics based on data mining and machine learning schemes. Nevertheless, the extension of infrastructures like OpenIoT with analytics functionalities is straightforward. In particular, such an extension entails the following steps:

- Integrating an analytics framework (such as the R project) in order to support the execution of machine learning functionalities.
- Implementing a data pre-processing (i.e. data preparation) layer, aiming at transforming the IoT data streams from the OpenIoT cloud to a format compatible with the analytics framework (e.g., R).
- Enhancing the concepts of the ontology in order to support additional devices, data streams and data analytics properties in a way that ensures the semantic unification of the various data streams to be produced prior to their integration to the analytics framework.

These steps provide a sound basis for advancing a Sensing-as-a-Service infrastructure to the IoT Analytics as a Service one. However, additional enhancements can be also implemented in order to ensure more scalable and high performance processing, through for example considering data storage, network latency and processing performance factors.

5.7 Conclusions

This chapter has focused on a special case of IoT/cloud integration, which entails the dynamic selection of sensors and the processing of their data towards a Sensing-as-a-Service paradigm. In addition to introducing the main principles of Sensing-as-a-Service, the chapter has also presented the practical aspects of this paradigm, based on a award-winning OpenIoT open source project. The latter provides technology and ease-to-use (visual) tools that enable the dynamic selection of virtual sensors from a cloud infrastructure and the subsequent processing of their data on the basis of functionalities that are provided by the SPARQL query language. The use of SPARQL as a data processing utility is enabled due to the semantic unification of the various IoT data streams, regardless of the (virtual) sensor that provides them. To this end, all IoT data streams are semantically annotated in order to comply with the same ontology. Overall, the OpenIoT project can be seen as a blueprint for implementing similar Sensing-as-a-Service systems.

The Sensing-as-a-Service paradigm can be also seen as a foundation for the implementation of IoT-Analytics-as-a-Service, through integrating more sophisticated data analytics capabilities over baseline Sensing-as-a-Service infrastructures. The latter provide a sound basis for IoT-Analytics-as-a-Service, since they offer the ever important data collection and semantic unification parts. We can expect a rise of IoT-Analytics-as-a-Service infrastructures in the near future, as enterprises are likely to seek opportunities for outsourcing complex the IoT analytics tasks to a cloud provider.

Acknowledgements

Part of this work has been carried out in the scope of the OpenIoT project (openiot.eu), which has been co-funded by the European Commission in the scope of the FP7 framework programme (contract No. 287305).

References

[1] Christian Vecchiola, Rajkumar Buyya, S. Thamarai Selvi. *Mastering Cloud Computing.* 1st Edition, Foundations and Applications Programming, Elsevier Print ISBN 9780124114548 Electronic ISBN 9780124095397.

[2] John Soldatos, M. Serrano and M. Hauswirth. *Convergence of Utility Computing with the Internet-of-Things*, International Workshop on Extending Seamlessly to the Internet of Things (esIoT), collocated at the IMIS-2012 International Conference, 4th–6th July, 2012, Palermo, Italy.

[3] John Soldatos, Nikos Kefalakis, et. al. *OpenIoT: Open Source Internet-of-Things in the Cloud.* Lecture Notes in Computer Science, invited paper, vol. 9001, (2015).

[4] Martin Serrano, John Soldatos. *IoT is More Than Just Connecting Devices: The OpenIoT Stack Explained*, IEEE Internet of Things Newsletter, September 8th, 2015.

[5] Martin Serrano, Hoan Nguyen Mau Quoc, Danh Le Phuoc, Manfred Hauswirth, John Soldatos, Nikos Kefalakis, Prem Prakash Jayaraman, Arkady B. Zaslavsky. *Defining the Stack for Service Delivery Models and Interoperability in the Internet of Things: A Practical Case With OpenIoT-VDK.* IEEE Journal on Selected Areas in Communications 33(4): 676–689 (2015).

[6] Karl Aberer, Manfred Hauswirth, Ali Salehi. *Infrastructure for Data Processing in Large-Scale Interconnected Sensor Networks.* MDM 2007: 198–205.

[7] Martin Serrano. *Applied Ontology Engineering in Cloud Services, Networks and Management Systems*. Springer publishers, March 2012. Hardcover, pp. 222 pages, ISBN-10: 1461422353, ISBN-13:978-1461422358.
[8] N. Kefalakis, S. Petris, C. Georgoulis, J. Soldatos. Open Source semantic web infrastructure for managing IoT resources in the Cloud. Book Chapter "Internet of Things: Principles and Paradigms", Elsevier Science, 2016, ISBN 978-0-12-809347-4.
[9] N. Kefalakis, J. Soldatos, A. Anagnostopoulos, and P. Dimitropoulos. A Visual Paradigm for IoT Solutions Development *in Interoperability and Open-Source Solutions for the Internet of Things*, Springer, 2015, pp. 26–45.
[10] J. Soldatos, N. Kefalakis, M. Serrano, and M. Hauswirth. Design principles for utility-driven services and cloud-based computing modelling for the Internet of Things. *Int. J. Web Grid Serv.*, vol. 10, no. 2, pp. 139–167, 2014.

6

A Review of Tools for IoT Semantics and Data Streaming Analytics

Martin Serrano and Amelie Gyrard

Insight Center for Data Analytics, National University of Ireland, Galway, Ireland

6.1 Introduction

The Internet of Things is having a fast adoption in the industry and with no doubt, in today's increasingly competitive innovation-driven marketplace, individuals need skills to transform the growing amount of industry, product, and customer (behavior) data into actionable information to support strategic and tactical decisions at the organisations. The Internet of Things (Devices) and the web are every day more and more crucial elements in this data generation, consolidating the need for interpreting data produced by those devices or "things", physical or virtual, connected to the Internet. In the other hand analytics is a time-consuming process and constantly redesigned in every application domain. The identification of methods and tools to avoid that data analytics over IoT Data is every time re-designed is a constant need. In this chapter, a review of tools for IoT data analytics is reviewed. We provide an overall vision on best practices and trends to perform analytics, we address innovative approaches using semantics to facilitate the integration of different sources of information [1].

The Semantic Web community plays a relevant role when interoperability of data and integration of results are required. The semantic analytics is an emerging initiative talking about Linked Open Data (LOD) reasoning, provides a vision on how to deduce meaningful information from IoT data, aiming to share the way to interpret data in a interoperable way to produce new knowledge [2]. Semantic analytics unifies different semantics technologies and analytic tools such as logic-based reasoning, machine learning, Linked Open Data (LOD), the main objective is to convert data into actionable knowledge, reasoning combines both semantic web technologies and reasoning approaches.

Collecting, transforming, interpreting data produced by devices "things" connected to the Internet/Web is a time-consuming process and constantly requires a redesign process for all applications that uses different data sources [3].

In this chapter, we analyse complementary research fields covering reasoning and semantic web approaches towards the common goal of enriching data within IoT. The involvement of semantics offers new opportunities and methods for production and discovery of information and also transforms information into actionable knowledge. The most common used methods are: 1) Linking Data, 2) Real-time and Linked Stream Processing, 3) Logic-based approaches, 4) Machine Learning based approaches, 5) Semantics-based distributed reasoning, and 6) Cross-domain recommender systems.

Ozpinar in 2014 [4] explained that resolving the meaning of data is a challenging problem and without processing it the data is invaluable. Pereira in 2014 [5] highlighted the necessity to interpret, analyse and understand sensor data to perform machine-to-machine communications. They classify six techniques such as supervised learning, unsupervised learning, rules, fuzzy logic, ontological reasoning and probabilistic reasoning in their survey dedicated to context-awareness for IoT. Further, they clearly explain pros and cons and sum up them in a table. According to their table, rule and ontology-based techniques contain few cons. Their shortcomings are to define manually rules which can be error prone and that there is no validation or quality checking. With such approaches, rules are only defined once in an interoperable manner. Pros concerning rule-based system are that rules are simple to define, easy to extend and require less computational resources. In semantic analytics and particularly 'Sensor-based Linked Open Rules' [6] will overcome these limitations, rules can be shared and reused and validated by domain experts. To deduce meaningful information from sensor data, the following main challenges to address are analysed: a) Real-time data, b) Scalability, c) Which machine learning algorithm should be apply for specific sensor datasets because there is a need to assist users in choosing the algorithm fitting their need, e) How to unify exiting systems and tools (e.g, S-LOR, LD4Sensors, KAT and LSM) since they are providing complementing approaches towards the same goal of enriching data, and f) How to extend KAT to assist experimenters to deal with machine learning and with real-time and to be compatible with the Stream Annotation Ontology (SAO).

Currently what is missing, are the methods to design innovative approaches for linked open data analytics. Recent approaches like Linked Open Reasoning (LOR), introduced in [2], inspired from the recent work in the context of European research projects. Linked Open Reasoning provides a solution to deduce meaningful information from IoT data and aims to share the way to interpret data in an interoperable way. This approach unifies different reasoning approaches such as logic-based reasoning and machine learning. Linked Open Reasoning combines both semantic web technologies and reasoning approaches. There is a vision that machine-learning approaches might not be necessary to interpret data produced by simple sensors. It will avoid the learning curve to deal with machine learning algorithms. The idea of "Linked Open Reasoning" (LOR) is an extension of our preliminary idea, Sensor-based Linked Open Rules (S-LOR) [6]. S-LOR is a dataset of interoperable IF THEN ELSE rules to deduce meaningful information from simple sensors such as thermometer.

6.2 Related Work

In this article, a complementary research covering reasoning and semantic web approaches towards the common goal of enriching data within IoT is presented and studied following the different approaches: 1) Linking Data, 2) Real-time and Linked Stream Processing, 3) Logic-based approaches, 4) Machine Learning based approaches, 5) Semantics-based distributed reasoning, and 6) Cross-domain recommender systems. We conclude by comparing different approaches and tools and highlighting the main limitations.

6.2.1 Linking Data

Karma is a data integration tool dealing with heterogeneous data such as XML, CSV, JSON, Web APIs, etc. based on ontologies and eases the publication of data semantically annotated with RDF [7]. This tool has been used to aggregate smart city data and semantically annotate data according to the KM4City ontology [8].

The LD4Sensors/inContext-Sensing is a tool that has been designed within the SPITFIRE EU project. This tool enriches sensor data with the Linked Data by using the Pachube API, the SPITFIRE ontology and the Silk tool to align datasets such as DBPedia, WordNet, Musicbrainz, DBLP, flickr wrappr and Geonames [9]. LD4Sensors provides JSON Web services, API and GUI to automate the annotation and linking process of sensor data. The semantic annotation is done with the Jena library and semantic data are stored using Jena TDB. LD4Sensors integrates a SPARQL endpoint to ease access to semantic sensor data. The semantic dataset and SPARQL endpoint are referenced on the dataset catalogue called DataHub. LD4Sensors provide linking but do not deal with real-time aspect nor interpret sensor data produced by devices by reusing domain-specific knowledge expertise.

6.2.2 Real-time & Linked Stream Processing

In recent years, a significant number of technologies that facilitate real-time and linked stream processing have also emerged. **Linked Stream Data** is an extension of the SPARQL query language and engine to deal with stream sensor data and enrich them with the Linked Open Data cloud [10]. **SPARQL is an RDF query language** and protocol produced by the W3C RDF Data Access Working Group (DAWG). SPARQL is extensively used in Semantic communities and was released as a W3C Recommendation in 2008. **C-SPARQL** was an earlier proposal of streaming SPARQL system [11]. Furthermore, the Continuous Query Evaluation over Linked Streams (CQELS) combines streaming capabilities and Linked Data [12, 13]. On top of this Le-Phuoc et al. [14] developed the SensorMasher and Linked Sensor Middleware (LSM) platforms in order to facilitate publishing of 'Linked Stream Data' and their use within other applications. In particular, they developed a user friendly interface to manage environmental semantic sensor networks. SPARQLStream is another novel approach for accessing and querying existing streaming data sources [15]. Specifically, SPARQLstream has been designed as an extension to the SPARQL 1.1 query language to deal with real-time sensor data [16].

6.2.3 Logic

Several mechanism and tools have also been developed in order to apply processing logic over streams. For example, Sensor-based Linked Open Rules (S-LOR) is an approach to share and reuse the rules to interpret IoT data, as explained in Section 6.3.3. It provides interoperable datasets of rules compliant with the Jena framework and inference engine. The rules have been written manually but are extracted from the Linked Open Vocabularies for Internet of Things (LOV4IoT) dataset [62, 63], an ontology/dataset/rule catalogue designed by domain experts in various applicative domains relevant for IoT such as healthcare, agriculture, smart home, smart city, etc.

Linked Edit Rules (LER) [17] is another recent approach similar to the Sensor-based Linked Open Rules (S-LOR) to share and reuse the rules associated to the data. This work has been not applied to the context of IoT. LER is more focused on checking consistency of data (e.g., a person's age cannot be negative, a man cannot be pregnant and an underage person cannot process a driving license). LER extends the RDF Data Cube data model by introducing the concept of EditRule. The implementation of LER is based on Stardogs rule reasoning to check obvious consistency.

Another relevant approach is provided by the **BASIL framework (Building APIs SImpLy)**, which combines REST principles and SPARQL endpoints in order to benefit from Web APIs and Linked Data [18]. BASIL reduces the learning curve of data consumers since they query web services exploiting SPARQL endpoints. The main benefit is that data consumers do not need to learn the SPARQL language and semantic web technologies.

6.2.4 Machine Learning

Machine learning is one of the most extended techniques in information systems, [19] and [20] are the earlier work to propose the idea to reason on semantic sensor data (e.g., to deduce potentially icy, blizzard, freezing concepts). The work described in [21] explains the idea of 'semantic perception' [22, 23] to interpret and reason on sensor data. This work developed an ontology of perception called IntellegO. A semantic-based approach to integrate abductive logic framework and Parsimonious Covering Theory (PCT) to integrate semantics in resource-constrained devices was also proposed. It explains that the development of background knowledge is a difficult task and out of the scope of this work. For this reason, recently, the LOV4IoT dataset has been designed to encourage the reuse of the domain knowledge expertise relevant for IoT. LOVIoT shows numerous challenges to automatically combine the background knowledge. IntellegO also illustrates that perception does not enable a straightforward formalization using logic-based reasoning. e.g., for simple sensors such as temperature or precipitation, logic-based reasoning is faster, flexible and easier for sharing. For more complex sensors such as accelerometers or ECG, logic-based reasoning is insufficient, and the uses of data mining approaches are unavoidable.

Beyond learning about the data, there are work (i.e. [24, 25]) introducing a Knowledge Acquisition Toolkit (KAT) to infer high-level abstractions from sensor data provided by gateways in order to reduce the traffic in network communications. KAT comprises three components: 1) An extension of Symbolic Aggregate Approximation (SAX) algorithm, called SensorSAX, 2) Abductive reasoning based on the Parsimonious Covering Theory (PCT), and 3) Temporal and spatial reasoning. It uses machine learning techniques (i.e. k-means clustering and Markov model methods) and Semantic Web Rule Language (SWRL) rule-based systems to add labels to the abstractions. KAT proposes the use of domain-specific background knowledge, which is not sufficient for Internet of Things, unless some another approach (e.g., the LOV4IoT dataset) is also integrated.

The work described in [26] employs the abductive model rather than inductive or deductive approaches to solve the incompleteness limitation due to missing observation information. The work is tested on real sensor data (i.e. temperature, light, sound, presence and power consumption). Their gateways support TinyOS, Contiki enabled devices and Oracle SunSpot nodes.

There are also approaches that emphasize the use of machine learning on sensor data [27]. This includes for example the use of decision trees and Bayesian network to analyze datasets comprising 16,578 measurements. The focus of the approach is on four kinds of sensor measurements: temperature, humidity, light and pressure. Furthermore, the dataset used has additional information such as weekday, hour interval, position of the window, number of computers working and number of people in the lab. An enrichment of sensor data with semantics has been also taken place [28, 29]. This enrichment provides context for sensor measurements, based on well-known ontologies such as Geonames for location, Geo WGS84 for coordinates, the W3C SSN ontology to describe sensors, the SWEET ontologies, as well as the W3C Time ontology. However, no need for IoT-related domain ontologies is expressed, while the need for semantic reasoners as a mean to infer new knowledge from sensor data is outlined [28].

The SemSense architecture [30] is one more approach to collect and then publish sensor data as Linked Data. Also, in [31], the authors collect data on the fly and then validate and link them with Linked Open Data (LOD) datasets. Devaraju et al. have also designed an ontology for weather events observed by sensors such as wind speed and visibility [32]. They are focused on blizzard related phenomena. They deduce high-level abstractions such as the types of snow (e.g., soft hail, snow, snow pellet, blizzard, winter storm, avalanche, flood, drought, and tornado). Such abstractions are deduced with rule-based reasoning, the implementation is based on Semantic Web Rule Language (SWRL)[1] and the Jess reasoning engine. The DUL ontology and the W3C SSN ontology are used. The approach is evaluated based on the Canadian Climate Archives database. In another work, Wang et al. explain that the SSN ontology "does not include modeling aspects for features of interest, units of measurement and domain knowledge that need to be associated with the

---

[1] https://www.w3.org/Submission/SWRL/

sensor data to support autonomous data communication, efficient reasoning and decision making" [33].

In recent years, there have also been efforts to interpret data produced by accelerometer, gyroscope, microphone, temperature and light sensors embedded in mobile phones [34]. These efforts use Hidden Markov Models (HMMs) and semantic web technologies to deduce activities. The rules are implemented as SPARQL queries. Moreover, Ramparany et al. introduced the need of a domain-specific automated reasoning system [35]. This work envisages that such a system could be based on Description Logic or Complex Event Processing (CEP) for interpreting IoT data. However, it does not propose a dataset with predefined rules that could be easily shared and reused by developers.

6.2.5 Semantic-based Distributed Reasoning

One of the early work on semantic-based distributed reasoning has been DRAGO, the Distributed Reasoning Architecture for a Galaxy of Ontologies, implemented as a peer-to-peer architecture [36]. The goal of DRAGO was to reason on distributed ontologies. Likewise, Kaonp2p has been designed to query over distributed ontologies [35]. Moreover, LarKC (Large Knowledge Collider) is another scalable platform for distributed reasoning [37]. Similarly, the Marvin framework is a scalable platform for parallel and distributing reasoning on RDF data [38]. Also, Schlicht et al. propose a peer-to-peer reasoning for interlinking ontologies [36]. These works outline the need to provide interoperable heterogeneous sensor-based rules and combine cross-domain ontologies and datasets in the context of IoT applications.

Abiteboul et al. have also approached the Web as a distributed knowledge base and proposed an automated reasoning over it [40]. This work demonstrated the importance of reusing sensor-based domain ontologies and rules. Also, WebPIE (Web-scale Parallel Inference Engine) is an inference engine for semantic web reasoning (OWL and RDFS) based on the Hadoop BigData platform [41]. WebPIE is scalable over 100 billion triples [42]. Another scalable system has been introduced by Coppens et al. [43] as an extension to the SPARQL query language to support distributed and remote reasoning. The relevant implementation of the system has been based on the Jena ARQ query engine. One more semantic reasoning framework for BigData has been introduced by Park et al. based on XOntology and SPARQL [44]. It uses the Hadoop platform, HDFS and MapReduce to deal with thousands of sensor data nodes.

Overall, it is noteworthy that none of the discussed distributed reasoning frameworks proposes and implements interoperable rules as a means of interpreting sensor data.

6.2.6 Cross-Domain Recommender Systems

Recently, cross-domain semantic and rule-based recommender systems have also been designed [45] or [46]. Such systems underline the importance of providing interoperable reasoning. Some works (e.g., [47, 48]) propose a domain-independent recommendation system to provide personalization services of different domains (tourism, movies, books). They incorporate semantics into a content-based system to improve the flexibility and the quality, a domain-based inference (side-ward propagation, upward propagation) for user's interests and a semantic similarity method is used to refine item-user matching algorithm. Such cross-domain recommender systems highlight the importance to provide a domain-independent reasoning.

At the same time, Hoxha et al. provide a cross-domain recommender system based on semantics and machine learning techniques (Markov logic) [45], while Tobias et al. provide a context-aware cross-domain recommender system. They exploit semantic web technologies and related tools such as DBpedia and the spreading activation algorithm [46]. These works underline the importance of a cross-domain reasoning that could also applied to sensor data.

6.2.7 Limitations of Existing Work

Most of the presented works have limitations when it comes to adding semantic capabilities for analytics in an IoT context. For example:

- LD4Sensors does not deal with real-time data and does not provide inference reasoning to deduce new information. However, datasets have been linked to get additional information.
- LSM does not integrate inference-reasoning engine to deduce new information.
- KAT has some usability limitations. Non-experts in machine learning have some difficulties to use this tool since they have to choose the algorithm without any assistance.
- S-LOR provides interoperable Jena rules. However, the same rules could be designed with SPARQL CONSTRUCT. Since SPARQL is a recommendation it would be better to share and reusing the rules according to SPARQL CONSTRUCT.

In Figure 6.1, we indicate the pros and cons of different approaches to enrich IoT data on the basis of: (A) Logic or rule-based reasoning, (B) Machine learning, (C) Linked Stream processing, (D) Reusing domain knowledge with

Methods	Pros	Cons
Logic/rule-based reasoning	- Simple rules - Adapted to simple sensors - Easy for beginners (learning & implementation) - Easier to combine rules	- Not adapted to complicated sensors - Heterogeneous rule languages and editors
Machine learning	- More elaborate results - Adapted to complicated sensors	- Need real datasets - Complicated for non-experts - Complicated for a 'sharing and reusing' approach
Linked Stream processing	- Real-time data - Scalability - Linked Data	- No real reasoning
Re-use domain knowledge (LOD, LOV, LOR)	- 'Sharing and reusing' approach	- Be familiar with semantics - Be familiar with ontology/instance matching tools - Not adapted to complicated sensors
Distributed Reasoning	- Scalability - Interoperability between systems	- Complicated for implementation
Recommendation systems	- Adapted to the user profile	- Complicated for non-experts - Need real datasets - Need user profile

Figure 6.1 Summary of existing approaches for IoT data enrichment.

Linked Open Data (LOD), Linked Open Vocabularies (LOV) and Linked Open Rules (LOR); (E) Distributed reasoning, and (F) Recommender systems.

Figure 6.2 depicts a classification of different tools according to the different approaches. Some of the existing works are based on machine learning algorithms. Usually, machine learning is employed when rule-based algorithms are infeasible. None of the existing works deals with the extraction, reuse and linking of rules already implemented in domain-specific projects. To deal with such limitations, there is a need to build a dataset of interoperable rules to reason on sensor data. To achieve this task, sensor data should be interoperable. This approach should be easy to be shared and reused by other projects. Since, SWRL rules are increasingly popular the approach will be based on this language. Further, sharing, reusing and combining SWRL rules will be typically easier than data mining algorithms.

6.3 Semantic Analytics

The semantic web community has designed open approaches for sharing and reusing open data by means of using Linked Data, Linked Vocabularies, and Linked Services as a first approach for enabling analytics. Inspired from

Methods / Tools	Logic/rule-based reasoning	Machine learning	Linked Stream processing	Re-use domain knowledge (LOD, LOV, LOR)	Distributed Reasoning	Recommen dation systems
'Semantic Sensor Web'	Yes	No	No	Yes	No	No
'Semantic Perception'	No	Yes	No	No	No	No
KAT	No	Yes	No	No	No	No
inContext-sensing	No	No	No	Yes	No	No
S-LOR	Yes	No	No	Yes	No	No
CQELS	No	No	Yes	Yes	No	No
SPARQLStream	No	No	Yes	Yes	No	No
WebPIE	No	No	No	No	Yes	No
DRAGO	No	No	No	No	Yes	No
Marvin	No	No	No	No	Yes	No
KaonP2P	No	No	No	No	Yes	No
LarKC	No	No	No	No	Yes	No

Figure 6.2 Classification of tools according to reasoning approaches.

the Semantic Web community semantic analytics plays a relevant role when interoperability of data and integration of results are required.

6.3.1 Architecture towards the Linked Open Reasoning

In this section, we describe and we attempt to unify reasoning approaches. Figure 6.3 presents a summary of the studied reasoning tools divided, for an easier understanding, in 3 layers and summarised as follow:

- The first layer (top) shown at the bottom provides API and web services to access to reasoning approaches. This layer provides access to simple reasoning services or complex services which are a composition of existing services.
- The second layer (middle) shows the generic reasoning approaches. We referenced, classified and analyzed the following different reasoning methods, including: (A) Machine Learning, which is a quite popular approach, yet it needs data to be integrated, which are not always available; (B) Real-time techniques, which are important when dealing with real-time data; (C) Linking techniques, which enable the enrichment of IoT data with background knowledge; (D) Complex Event Processing (CEP) approaches, which apply inference based reasoning in order to extract or deduce new information from IoT data; (E) Sharing and

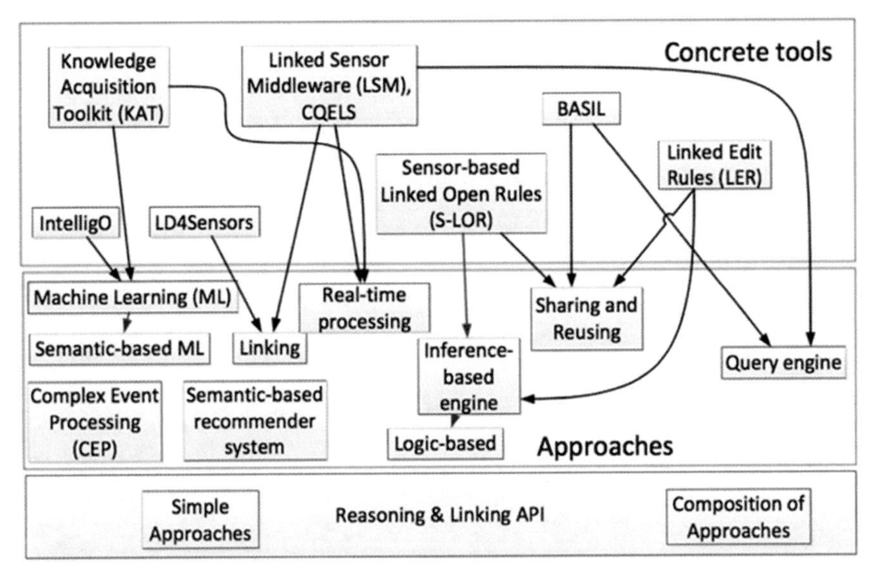

Figure 6.3 Reasoning main operations.

reusing approaches, which enable deduction of new information from data produced by Internet Connected Objects (ICOs) in a way similar to Linked Open Data.

- The third layer (bottom) indicates the concrete tools that we could reuse and unify to interpret data. Such tools are: (A) the Knowledge Acquisition Toolkit (KAT) which is a machine-learning approach dealing with real-time data; (B) The Linked Sensor Middleware (LSM), which deals with real-time data and enables linking between heterogeneous datasets; (C) IntelligO, which is a machine-learning approach using the Parsimonious Covering Theory (PCT); (D) Linked Data for Sensors (LD4Sensors), which enables linking of sensor datasets and (E) Sensor-based Linked Open Rules (S-LOR) which is a rule-based reasoning engine and innovative approach to share and reuse interoperable deductive rules in order to infer new knowledge from IoT data.

6.3.2 The Workflow to Process IoT Data

Figure 6.4 illustrates different processes and steps required to combine data from heterogeneous sources and to build innovative and interoperable applications. The figure illustrates the SEG 3.0 methodology [2] that is extended and used for interoperability but also for semantic analytics. In this book chapter, we are mainly focused on the reasoning layer. It comprises the following steps:

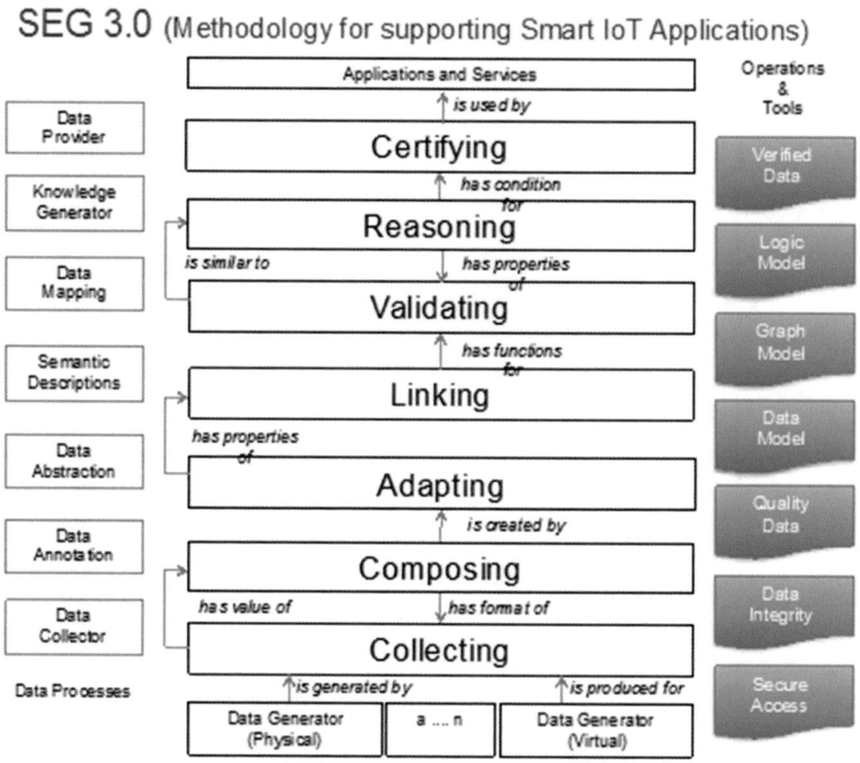

Figure 6.4 IoT process defined by SEG 3.0 methodology.

- **Composition of IoT data**, enabling the unification of heterogeneous data coming from different IoT sources and particularly re-using different data formats (e.g., CSV, Excel) or different terms (e.g., temp or temperature). This activity requires a common dictionary to unify terms employed to describe data. For example, the composing layer could return the SenML format to describe sensor data [49].
- **IoT Data Modeling**, which enables the annotation of data with semantic web technologies (e.g., RDF, RDFS and OWL). This step employs models, vocabularies and ontologies to unify data, which is prerequisite for the next steps. The M3 ontology is used to unify semantic sensor data [50].
- **Linking IoT data domains**, enabling the enrichment of data with metadata from other RDF datasets to get additional information. It exploits the idea of Linked Data and Linked Vocabularies for IoT applications.

- **Reasoning over IoT Streaming Data**, which enables the updating of the database/triple store with additional triples for instance by using a reasoning engine (e.g., Jena rule-based inference engine) to infer high level abstraction from data. It exploits the idea of Linked Rules.
- **Querying IoT Data**, which enables the querying of RDF datasets through the SPARQL language based on ontologies used in the previous steps. It is an essential step to get data and build end-users services/applications.
- **IoT Services activation and control**, which enables end-users to access smarter data. The data is available through interoperable APIs or web services (e.g., RESTful web services). Such web services returns the result provided by the SPARQL query engine.
- **Composition of IoT services**, which enables the development of complex applications based on the composition of several services. It can be achieved through the use of web services or semantic web services.

The SEG 3.0 methodology supports the vision of semantic interoperability from data to end-users applications, which is inspired from the 'sharing and reusing' based approach as depicted in Figure 6.4. The realization of the vision is based on a combination of several of the concepts that have been already presented including: 1) Linked Open Data (LOD); 2) Linked Open Vocabularies (LOV); 3) Linked Open Rules/Reasoning (LOR); and 4) Linked Open Services (LOS).

In the following paragraphs, we extend and apply these approaches to IoT and smart cities. Note that Linked Open Data (LOD) is an approach to share and reuse the data [51, 52]. However, previous works on 'Linked Sensor Data' [53, 54] do not provide any tools for visualizing or navigating through IoT datasets. For this reason, a Linked Open Data Cloud for Internet of Things (CLOuDIoT) infrastructure to share and reuse data produced by sensors is being implemented.

Linked Open Vocabularies (LOV) is an approach to share and reuse the models/vocabularies/ontologies [55]. LOV did not reference any IoT ontologies. For this reason, we have also designed the Linked Open Vocabularies for Internet of Things (LOV4IoT) [62, 63], a dataset of almost 300 ontology-based IoT projects referencing and classifying: 1) IoT applicative domains, 2) Sensors used, 3) Ontology status (e.g., shared online, best practices followed), 4) Reasoning used to infer high level abstraction, and 5) Research articles related to the project. This dataset contains the background knowledge required to add value to the data produced by Internet Connected Objects (ICOs).

6.3.3 Sensor-based Linked Open Rules (S-LOR)

Linked Open Reasoning (LOR) is an approach for sharing and reusing the ways to interpret the data and to deduce new information (e.g., machine learning algorithm used, reusing rules already designed by domain experts). To this end, LOR can be extended towards using semantics Sensor-based Linked Open Rules (S-LOR), a dataset of interoperable rules (e.g., if-then-else rules) used to interpret data produced by sensors [56]. Such rules are executed with an inference engine (e.g., Jena) that updates the triple store with additional triples. For example, the rule can be if the body temperature is greater than 38 degree Celsius than fever. In this example, the triple store will be updated with this high level abstraction 'fever'. The approach is inspired from the idea of 'Linked Rules' [57] that provides a language to interchange semantic rules but not the idea of reusing existing rules.

6.4 Tools & Platforms

Data analytics is a complex activity that requires examining raw data with the purpose of drawing conclusions. In IoT, the combination of different data types, in nature and in format, makes this practice more complex. Data analytics is extensively used in science to verify or eliminate existing models or theories, analytics is also used in many domains to allow companies and organization to make better business decisions. Data analytics focuses on inference, the process of generate conclusion(s) based solely on what is already known by the researcher.

On the other hand, semantic analytics is an advanced technique that uses the normalisation of data to a one particular format with the advantage of data alignment and interoperability. This allows the generation of more information. Necessary steps for semantic analytics, along with related tools are presented in the following paragraphs.

6.4.1 Semantic Modeling and Validation Tools

A variety of semantic modeling tools have recently emerged and are already used in the scope of IoT applications. For example:

- **HyperThing**[2] is a semantic web URI validator, which determines whether a URI identies a Real World Object or a Web document resource. It checks whether the URIs publishing method follows the W3C hash URIs and 303 URI practices. It can also be used to check the validity of the

[2] http://www.hyperthing.org

chains of the redirection between the Real World Object URIs and Document URIs, in order to prevent the data publisher mistakenly redirect.
- **NeON**[3] is a methodology for Ontology Engineering in a networked world.
- **OWL Validator** is another semantic validation project that accepts ontologies written in RDF/XML, OWL/XML, OWL Functional Syntax, Manchester OWL Syntax, OBO Syntax, and KRSS Syntax.
- **OQuaRE**[4] is a square-based approach for evaluating the quality of ontologies. OQuaRE covers two main processes: software quality requirement specifications and software quality evaluation.
- **OntoClean**[5] provides a definition of metaproperties that help with the construction of ontology language descriptions of problem domains.
- **OnToology**[6] is a system to automate part of the collaborative ontology development process. OnToology works surveying a repository with an OWL file, produce diagrams, a complete documentation and do validation based on common pitfalls.
- **Oops**[7] helps ontology designers detect some of the most common pitfalls appearing within ontology developments in particular when: (a) the domain or range of a relationship is defined as the intersection of two or more classes; (b) no naming convention is used within the identifiers of the ontology elements; and (c) a cycle between two classes in the hierarchy is included in the ontology.
- **Ontocheck**[8] is a tool for verifying ontology naming conventions and metadata completeness following cardinality checks on mandatory and obligatory annotation properties and reviewing naming conventions via lexical analysis and labeling enforcement.
- **OntoAPI**[9] is a consumer API that can process the response of an ontology evaluation web service provider.
- **OntoMetric**[10] is a method to choose the appropriate ontology.
- **Prefix**[11] simplifies the tedious task of any RDF developer, by remembering and looking up URI prefixes.

[3]http://neon-toolkit.org/wiki/Main Page
[4]http://miuras.inf.um.es/oquarewiki/index.php5/MainPage
[5]http://www.ontoclean.org/
[6]http://ontoology.linkeddata.es
[7]http://oops.linkeddata.es
[8]http://www2.imbi.uni-freiburg.de/ontology/OntoCheck/
[9]https://sourceforge.net/projects/drontoapi/
[10]http://oa.upm.es/6467/
[11]http://prefix.cc/

- **Vapour**[12] is a linked Data Validator in the form of a scripting approach to debug content. Vapour facilitates the task of testing the results of content negotiation on a vocabulary.
- **Vocab**[13] is an open source project that allows RDF developers to look up and search for Linked Data vocabularies. Developers can search URIs with arbitrary queries or look up specific URIs.
- **The W3C RDF Validator**[14] is an online service for checking and visualizing your RDF documents, W3C RDF validator is based on Another RDF Parser.

6.4.2 Data Reasoning

There are also numerous data reasoners that enable knowledge generation and activation. However not all of them are in a mature stage nor serve the same purpose. By means of its level of complexity to configure actionable data in the IoT, reasoners can be catalogued not only by their linkage and discovery mechanisms, but also on the basis of their usability in the area. The following selection is a comprehensive list of reasoners that can be used in the scope of IoT streaming and IoT analytics applications:

- **CEL DL (Description Logic)**[15] is a reasoner which implements a polynomial-time algorithm. The supported description logic (EL+) offers a selected set of expressive means that are tailored towards the formulation of domain-specific ontologies. CEL's main reasoning task is the computation of the subsumption hierarchy induced by EL+ ontologies.
- **Euler**[16] is an inference engine supporting logic-based proofs. It is a backward-chaining reasoner enhanced with Euler path detection. It has implementations in Java, C#, Python, JavaScript and Prolog. In conjunction with N3 it is interoperable with W3C Cwm.
- **FaCT++**[17] is the new generation of the well-known FaCT OWL-DL reasoner. FaCT++ uses the established FaCT algorithms, but with a different internal architecture.
- **HermiT**[18] is a highly efficient OWL reasoner. HermiT is a reasoner for ontologies written using the Web Ontology Language (OWL). HermiT

[12]http://linkeddata.uriburner.com:8000/vapour
[13]http://vocab.cc
[14]http://www.w3.org/RDF/Validator/
[15]https://lat.inf.tu-dresden.de/systems/cel/
[16]https://www.w3.org/2001/sw/wiki/Euler
[17]http://owl.man.ac.uk/factplusplus/
[18]http://www.hermit-reasoner.com

is based on a novel "hypertableau" calculus which provides much more efficient reasoning than any previously-known algorithm.

- **JESS (Java Expert System Shell)**[19] is in the form of a Jena inference implementation with rule engine and scripting environment written entirely in Java^{TM}.
- **Jena Eyeball**[20] is a command-line semantics validator for checking RDF/OWL common problems. Eyeball is a Jena-based tool for checking RDF models (including OWL) for common problems. It is user-extensible using plugins.
- **Kaon2**[21] is an OntoBroker designed for managing ontologies. KAON2 is a successor to the KAON project often referred to as KAON1. The main difference to KAON1 is the supported ontology language: KAON1 used a proprietary extension of RDFS, whereas KAON2 is based on OWL-DL and F-Logic.
- **Nools**[22] is a RETE based rule engine written entirely in javascript. When using nools tool, a flow which acts as a container for rules that can later be used to get a session.
- **OWLlink API**[23] is designed to access remote reasoners. OWLlink API has a Java interface for the OWLlink protocol on top of the Java-based OWL API. The OWLlink API enables OWL API-based applications to access remote reasoners (so-called OWLlink servers), and it turns any OWL API aware reasoner into an OWLlink server.
- **Pellet**[24] is an open-source Java based OWL 2 reasoner, It can be used in conjunction with both Jena and OWL API libraries.
- **Racer Pro**[25] is an OWL reasoner tool that can perform reasoning tasks. Racer pro has an inference server for the Semantic Web.
- **RIF4j**[26] is a reasoning engine for RIF-BLD that provides a Java object model for RIF-BLD and supports the parsing and serialization of RIF-BLD formulas. Furthermore, it provides a prototype implementation of a RIF-BLD consumer based on the Datalog engine IRIS.

---

[19] http://www.jessrules.com

[20] http://jena.apache.org/documentation/tools/eyeballgetting-started.html

[21] http://kaon2.semanticweb.org

[22] http://c2fo.io/nools

[23] http://owllink-owlapi.sourceforge.net

[24] https://www.w3.org/2001/sw/wiki/Pellet

[25] https://www.w3.org/2001/sw/wiki/RacerPro

[26] http://rif4j.sourceforge.net

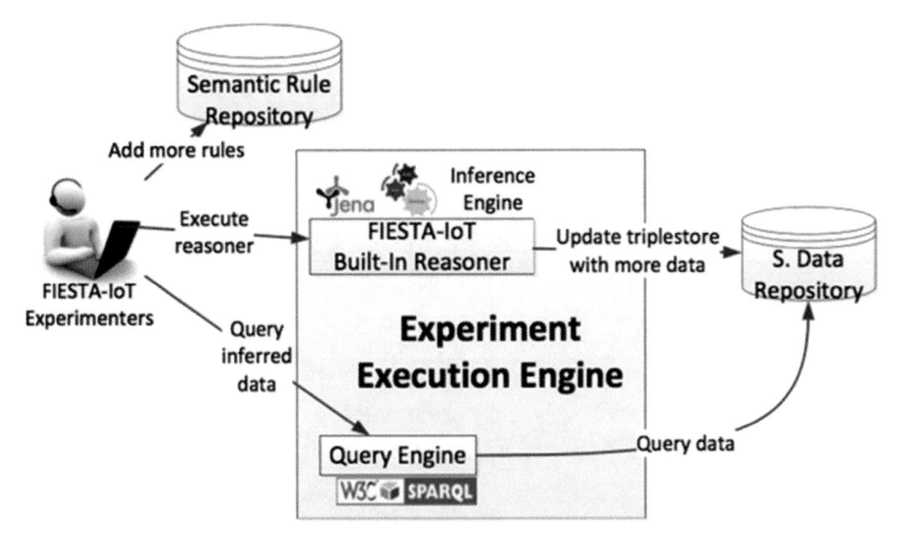

Figure 6.5 IoT reasoning data framework within FIESTA-IoT.

6.5 A Practical Use Case

Federated Interoperable Semantic IoT/cloud Testbeds and Applications (FIESTA-IoT)[27] is an EU project (funded in the context of the H2020 framework), which focuses on integrating IoT platforms, testbeds, data and associated silo applications. FIESTA-IoT opens up new opportunities in the development and deployment of experiments that exploit data and capabilities from multiple geographically and administratively dispersed IT testbeds. The project employs semantic (ontology) modeling as a mechanism to associate different domains and beyond that discover relationships amongst the information.

Figure 6.5 shows the designed FIESTA-IoT reasoning engine approach that by design will be used by experimenters/users of the platform. Based on our analysis of the literature, a logic-based/rule-based reasoning is used. Experimenters can interact with the reasoning engine as follows:

- Increasing the actionable knowledge by contributing to the Semantic Rule Repository, a dataset of interoperable rules. Such rules are IF THEN ELSE rules.

[27]http://fiesta-iot.eu/

- Executing the reasoning engine to infer additional information. Once executed, the reasoning engine updates the triple-store with additional triples (e.g., high level information).
- Querying inferred data, by executing a query engine that interacts with the triple store, called Semantic Data Repository.
- Implementing the rules as Jena rules since we used the Jena framework to build semantic web applications. Moreover, Jena provides an inference engine to easily execute the Jena rules and deduce additional information. To enhance interoperability, Jena rules can be designed as SPARQL CONSTUCT rules.

An overview of the FIESTA-IoT system is provided in Figure 6.5.

6.6 Conclusions

In this chapter, a summary of complementary research fields covering reasoning and semantic web approaches towards the common goal of enriching data within the IoT domain has been presented. The different aspects around semantic analytics like Linking Data, Real-time and Linked Stream Processing, Logic-based approaches, Machine Learning based approaches, semantics-based distributed reasoning, and cross-domain recommender systems, have been summarized and discussed.

The presented approaches and tools are able to deduce meaningful information from IoT data, based on the combination and integration of best practices from the literature. This approach is currently applied in the context of the H2020 FIESTA-IoT project. It leverages a combination of concepts and tools associated with Linked Open Data, Linked Open Vocabularies, Linked Open Services and Linked Open Reasoning.

Acknowledgement

This work is partially funded by the "Federated Interoperable Semantic IoT/cloud Testbeds and Applications" (FIESTA-IoT) project, which is funded by the European Commission in the scope of its Horizon 2020 Programme (Grant Agreement No. CNECT-ICT-643943). The literature survey regarding analytics tools is extended from the authors' past work particularly from the PhD thesis [61] under the supervision of Prof. Christian Bonnet and Dr. Karima Boudaoud.

References

[1] M. Serrano, et al. Cousin, "*Semantic Interoperability: Research Challenges, Best Practices, Solutions and Next Steps*", IERC AC4 Manifesto," 2014.

[2] A. Gyrard and M. Serrano, "*Connected Smart Cities: Interoperability with SEG 3.0 for the Internet of Things*" Proc. 30th IEEE Int'l Conf. Advanced Information Networking and Applications Workshops, 2016.

[3] M. Serrano et al., "*Defining the Stack for Service Delivery Models and Interoperability in the Internet of Things: A Practical Case With OpenIoT-VDK*," IEEE J. Selected Areas in Comm., vol. 33, no. 4, 2015, pp. 676–689.

[4] M. Özpinar, "*A flexible semantic service composition framework for pervasive computing environments*" Ph.D. dissertation, Moddle East Technical University, 2014.

[5] C. Perera, A. Zaslavsky, P. Christen, and D. Georgakopoulos, "*Context aware computing for the internet of things: A survey*", Communications Surveys & Tutorials, IEEE, vol. 16, no. 1, pp. 414–454, 2014.

[6] A. Gyrard, C. Bonnet, and K. Boudaoud, "*Helping IoT application developers with sensor-based linked open rules*," in SSN 2014, 7th International Workshop on Semantic Sensor Networks in conjunction with the 13th International Semantic Web Conference (ISWC 2014), 19–23 October 2014, Riva Del Garda, Italy, 10 2014.

[7] C. A. Knoblock, P. Szekely, J. L. Ambite, A. Goel, S. Gupta, K. Lerman, M. Muslea, M. Taheriyan, and P. Mallick, "Semi-automatically mapping structured sources into the semantic web," in The Semantic Web: Research and Applications. Springer, 2012, pp. 375–390.

[8] P. Bellini, M. Benigni, R. Billero, P. Nesi, and N. Rauch, "Km4city ontology building vs data harvesting and cleaning for smart-city services," Journal of Visual Languages & Computing, vol. 25, no. 6, pp. 827–839, 2014.

[9] M. Leggieri, A. Passant, and M. Hauswirth, "incontext-sensing: Lod augmented sensor data?" in Proceedings of the 10th International Semantic Web Conference (ISWC 2011). Citeseer, 2011.

[10] J. F. Sequeda and O. Corcho, "Linked stream data: A position paper," 2009.

[11] D. F. Barbieri, D. Braga, S. Ceri, E. Della Valle, and M. Grossniklaus, "C-sparql: Sparql for continuous querying," in Proceedings

of the 18th international conference on World wide web. ACM, 2009, pp. 1061–1062.

[12] D. Le Phuoc, "A native and adaptive approach for linked stream data processing," Ph.D. dissertation, 2013.

[13] D. Le-Phuoc, M. Dao-Tran, J. X. Parreira, and M. Hauswirth, "A native and adaptive approach for unified processing of linked streams and linked data," in The Semantic Web-ISWC 2011. Springer, 2011, pp. 370–388.

[14] D. Phuoc and M. Hauswirth, "Linked open data in sensor data mashups," 2009.

[15] J.-P. Calbimonte, O. Corcho, and A. J. Gray, "Enabling ontology-based access to streaming data sources," in The Semantic Web-ISWC 2010. Springer, 2010, pp. 96–111.

[16] J.-P. Calbimonte, "Ontology-based access to sensor data streams," Ph.D. dissertation, Informatica, 2013.

[17] A. Meroño-Peñuela, C. Guéret, and S. Schlobach, "Linked edit rules: A web friendly way of checking quality of rdf data cubes."

[18] E. Daga, L. Panziera, and C. Pedrinaci, "A basilar approach for building web apis on top of sparql endpoints," 2015.

[19] A. Sheth, C. Henson, and S. Sahoo, "Semantic sensor web," Internet Computing, IEEE, vol. 12, no. 4, pp. 78–83, 2008.

[20] W. Wei and P. Barnaghi, "Semantic annotation and reasoning for sensor data," Smart Sensing and Context, pp. 66–76, 2009.

[21] C. A. Henson, "A semantics-based approach to machine perception," Ph.D. dissertation, Wright State University, 2013.

[22] C. Henson, A. Sheth, and K. Thirunarayan, "Semantic perception: Converting sensory observations to abstractions," Internet Computing, IEEE, vol. 16, no. 2, pp. 26–34, 2012.

[23] P. Barnaghi, F. Ganz, C. Henson, and A. Sheth, "Computing perception from sensor data," 2012.

[24] F. Ganz, P. Barnaghi, and F. Carrez, "Information abstraction for heterogeneous real world internet data," Sensors Journal, IEEE, vol. 13, no. 10, pp. 3793–3805, Oct 2013.

[25] F. Ganz, P. Barnaghi, and F. Carrez, "Automated semantic knowledge acquisition from sensor data," 2014.

[26] F. Ganz, "Intelligent communication and information processing for cyber-physical data," Ph.D. dissertation, University of Surrey, 04 2014.

[27] A. Moraru, M. Pesko, M. Porcius, C. Fortuna, and D. Mladenic, "Using machine learning on sensor data," CIT. Journal of Computing and Information Technology, vol. 18, no. 4, pp. 341–347, 2010.

[28] A. Moraru, "Enrichment of sensor descriptions and measuremebts using semantic technologies," Master's thesis, 2011.

[29] A. Moraru and D. Mladenić, "A framework for semantic enrichment of sensor data," CIT. Journal of Computing and Information Technology, vol. 20, no. 3, pp. 167–173, 2012.

[30] A. Moraru, D. Mladenic, M. Vucnik, M. Porcius, C. Fortuna, and M. Mohorcic, "Exposing real world information for the web of things," in Proceedings of the 8th International Workshop on Information Integration on the Web: in conjunction with WWW 2011. ACM, 2011, p. 6.

[31] L. Bradeško, A. Moraru, B. Fortuna, C. Fortuna, and D. Mladenić, "A framework for acquiring semantic sensor descriptions (short paper)," Semantic Sensor Networks, p. 97, 2012.

[32] A. Devaraju and T. Kauppinen, "Sensors tell more than they sense: Modeling and reasoning about sensor observations for understanding weather events," International Journal of Sensors Wireless Communications and Control, vol. 2, no. 1, pp. 14–26, 2012.

[33] W. Wang, S. De, G. Cassar, and K. Moessner, "Knowledge representation in the internet of things: semantic modelling and its applications," Automatika-Journal for Control, Measurement, Electronics, Computing and Communications, vol. 54, no. 4, 2013.

[34] B. Boshoven and P. van Bommel, "Personalized life reasoning," Master's thesis, 2014.

[35] F. Ramparany, F. G. Marquez, J. Soriano, and T. Elsaleh, "Handling smart environment devices, data and services at the semantic level with the fiware core platform," in BigData (Big Data), 2014 IEEE International Conference on. IEEE, 2014, pp. 14–20.

[36] L. Serafini and A. Tamilin, "Drago: Distributed reasoning architecture for the semantic web," in The Semantic Web: Research and Applications. Springer, 2005, pp. 361–376.

[37] D. Fensel, F. van Harmelen, B. Andersson, P. Brennan, H. Cunningham, E. Della Valle, F. Fischer, Z. Huang, A. Kiryakov, T.-I. Lee et al., "Towards larkc: a platform for web-scale reasoning," in Semantic Computing, 2008 IEEE International Conference on. IEEE, 2008, pp. 524–529.

[38] E. Oren, S. Kotoulas, G. Anadiotis, R. Siebes, A. ten Teije, and F. van Harmelen, "Marvin: Distributed reasoning over large-scale semantic web data," Web Semantics: Science, Services and Agents on the World Wide Web, vol. 7, no. 4, pp. 305–316, 2009.

[39] A. Schlicht and H. Stuckenschmidt, "Peer-to-peer reasoning for interlinked ontologies," International Journal of Semantic Computing, vol. 4, no. 01, pp. 27–58, 2010.

[40] S. Abiteboul, E. Antoine, and J. Stoyanovich, "Viewing the web as a distributed knowledge base," in Data Engineering (ICDE), 2012 IEEE 28th International Conference on. IEEE, 2012, pp. 1–4.

[41] J. Urbani, S. Kotoulas, J. Maassen, F. Van Harmelen, and H. Bal, "Webpie: A web-scale parallel inference engine using mapreduce," Web Semantics: Science, Services and Agents on the World Wide Web, vol. 10, pp. 59–75, 2012.

[42] J. Urbani, "On web-scale reasoning," Ph.D. dissertation, 2013.

[43] S. Coppens, M. Vander Sande, R. Verborgh, E. Mannens, and R. Van de Walle, "Reasoning over sparql," in Proceedings of the 6th Workshop on Linked Data on the Web, 2013.

[44] K. Park, Y. Kim, and J. Chang, "Semantic reasoning with contextual ontologies on sensor cloud environment," International Journal of Distributed Sensor Networks, vol. 2014, 2014.

[45] J. Hoxha, "Cross-domain recommendations based on semantically-enhanced user web behavior," Ph.D. dissertation, Karlsruhe, Karlsruher Institut für Technologie (KIT), Diss., 2014, 2014.

[46] I. F. Tobías, "A semantic-based framework for building cross-domain networks: Application to item recommendation," Ph.D. dissertation, 2013.

[47] V. Codina and L. Ceccaroni, "Taking advantage of semantics in recommendation systems," in Proceedings of the 2010 Conference on Artificial Intelligence Research and Development: Proceedings of the 13th International Conference of the Catalan Association for Artificial Intelligence. Amsterdam, The Netherlands, The Netherlands: IOS Press, 2010, pp. 163–172. [Online]. Available: http://dl.acm.org/citation.cfm?id=1893268.1893291

[48] V. Codina and L. Ceccaroni, "A recommendation system for the semantic web," in Distributed Computing and Artificial Intelligence. Springer, 2010, pp. 45–52.

[49] J. Arkko, "Network working group c. jennings internet-draft cisco intended status: Standards track z. shelby expires: January 18, 2013 sensinode," 2012, media Type for Sensor Markup Language (SENML), draft-jennings-senml-09 (work in progress).

[50] A. Gyrard and C. Bonnet, "A unified language to describe M2M/IoT data," M3 2015.

[51] C. Bizer, T. Heath, and T. Berners-Lee, "Linked data-the story so far," International Journal on Semantic Web and Information Systems (IJSWIS), vol. 5, no. 3, pp. 1–22, 2009, http://www.w3.org/DesignIssues/LinkedData.html
[52] L. R. e. M. H. David Wood, Marsha Zaidman, Linked Data. Structured Data on the Web, 2014.
[53] H. Patni, C. Henson, and A. Sheth, "Linked sensor data," in Collaborative Technologies and Systems (CTS), 2010 International Symposium on. IEEE, 2010, pp. 362–370.
[54] P. Barnaghi, M. Presser, and K. Moessner, "Publishing linked sensor data," in CEUR Workshop Proceedings: Proceedings of the 3rd International Workshop on Semantic Sensor Networks (SSN), Organised in conjunction with the International Semantic Web Conference, vol. 668, 2010.
[55] P.-Y. Vandenbussche, G. A. Atemezing, M. Poveda-Villalón, and B. Vatant, "Lov: a gateway to reusable semantic vocabularies on the web," Semantic Web Journal, 2015.
[56] A. Gyrard, C. Bonnet, and K. Boudaoud, "Helping IoT application developers with sensor-based linked open rules," in SSN 2014, 7th International Workshop on Semantic Sensor Networks in conjunction with the 13th International Semantic Web Conference (ISWC 2014), 19–23 October 2014, Riva Del Garda, Italy, 10 2014.
[57] A. Khandelwal, I. Jacobi, and L. Kagal, "Linked rules: principles for rule reuse on the web," in Web Reasoning and Rule Systems. Springer, 2011, pp. 108–123.
[58] S. Speiser and A. Harth, "Towards linked data services," in Proceedings of the 9th International Semantic Web Conference (ISWC). Citeseer, 2010.
[59] S. Speiser and A. Harth, "Taking the lids off data silos," in Proceedings of the 6th International Conference on Semantic Systems. ACM, 2010, p. 44.
[60] B. Norton and R. Krummenacher, "Consuming dynamic linked data." in COLD, 2010.
[61] Designing Cross-Domain Semantic Web of Things Applications Amelie Gyrard's PhD Thesis, Eurecom, Sophia Antipolis, 24 April 2015.
[62] Reusing and Unifying Background Knowledge for Internet of Things with LOV4IoT 4th International Conference on Future Internet of Things

and Cloud (FiCloud 2016), 22–24 August 2016, Vienna, Austria Amelie Gyrard, Ghislain Atemezing, Christian Bonnet, Karima Boudaoud and Martin Serrano.

[63] LOV4IoT: A second life for ontology-based domain knowledge to build Semantic Web of Things applications 4th International Conference on Future Internet of Things and Cloud (FiCloud 2016), 22–24 August 2016, Vienna, Austria Amelie Gyrard, Christian Bonnet, Karima Boudaoud and Martin Serrano.

PART II

IoT Analytics Applications and Case Studies

7

Data Analytics in Smart Buildings

M. Victoria Moreno, Fernando Terroso-Sáenz, Aurora González-Vidal and Antonio F. Skarmeta

Department of Information and Communications Engineering, University of Murcia, 30100 Spain

7.1 Introduction

Cities are becoming more and more of a focal point for our economies and societies at large, particularly because of on-going urbanisation, and the trend towards increasingly knowledge-intensive economies as well as their growing share of resource consumption and emissions. To meet public policy objectives under these circumstances, cities need to change and develop, but in times of tight budgets this change needs to be achieved in a smart way: our cities need to become "smart cities". In order to follow the policy of the decarbonisation of Europe's economy in line with the EU 20/20/20 energy and climate goals, today's ICT, energy (use), transport systems and infrastructures have to drastically change. The EU needs to shift to sustainable production and use of energy, to sustainable mobility, and sustainable ICT infrastructures and services. Cities and urban communities play a crucial role in this process. Three quarters of our citizens live in urban areas, consuming 70%[1] of the EU's overall energy consumption and emitting roughly the same share of Green House Gas (GHG). Of that, buildings and transport represent the lion's share. Within the worldwide perspective of energy efficiency, it is important to highlight that buildings are responsible for 40% of total EU energy consumption and generate 36% of GHG [1]. This indicates the need to achieve energy-efficient buildings to reduce their CO2 emissions and their energy consumption.

Moreover, the building environment affects the quality of life and work of all citizens. Thus, buildings must be capable of not only providing mechanisms

[1]Source: European Commission 2013.

to minimize their energy consumption (even integrating their own energy sources to ensure their energy sustainability), but also of improving occupant experience and productivity. In this chapter, we analyse the important role that buildings represent in terms of their energy performance at city level and, even, at world level, where they represent an important factor for the energy sustainability of the planet. Analysis of the energy efficiency of the built environment has received growing attention in the last decade [2–4]. Various approaches have addressed energy efficiency of buildings using predictive modelling of energy consumption based on usage profiles, climate data and building characteristics. On the other hand, studies have demonstrated the impact of displaying public information to occupants and its effect in modifying individual behaviour in order to obtain energy savings [5, 6]. Nevertheless, most of the approaches proposed to date only provide partial solutions to the overall problem of energy efficiency in buildings, where different factors are involved in a holistic way, but which, until now, have been addressed separately or even neglected by previous proposals. This division is frequently due to the uncertainty and lack of data and inputs included in the management processes, so that analysis of how energy in buildings is consumed is incomplete. In other words, a more integral vision is required to provide accurate models of the energy consumed in buildings [7].

The need for the robust characterization of energy use in buildings has gained attention in light of the growing number of projects and developments addressing this topic. Although much interest has been put into smart building technologies, the research area of using real-time information has not been fully exploited. In order to obtain an accurate simulation model, a detailed representation of the building structure and its subsystems is required, although it is the integration of all these pieces that requires the most significant effort.

The integration and development of systems based on ICT and, more specifically, the IoT [8], are important enablers of a broad range of applications, both for industries and the general population, helping make smart buildings a reality. IoT permits the interaction between smart things and the effective integration of real world information and knowledge in the digital world. Smart (mobile) things endowed with sensing and interaction capabilities or identification technologies (such as RFID) provide the means to capture information about the real world in much more detail than ever before.

Regarding this real-world data extraction, the great adoption of personal handheld devices, like smartphones, has enabled the crowdsensing paradigm as a prominent mechanism to capture a wide range of (mobile) data [9]. Unlike other sensing approaches, in this case the collected data is directly generated by the users' personal contrivances, so it can be a useful solution for soliciting feedback from a sheer number of people in an explicit or implicit manner. From a smart building perspective, such feedback provides information about its occupants' preferences and habits that could be considered in order to come up with customized energy-efficiency solutions.

Nevertheless, challenges related to: (1) the management of the huge amount of data provided in real-time by a large number of IoT and crowd based devices deployed, (2) the interoperability among different ICT, and (3) the integration of many proprietary protocols and communication standards that coexist in the ICT market applicable to buildings (such as heating, cooling and air conditioning machines), need to be faced before flexible and scalable solutions based on the IoT paradigm can be offered.

The structure of the present chapter is as follows: Section 7.2 describes the key issues involved in energy efficiency in buildings. Among these issues, relevant parameters affecting energy consumed in buildings are described and proposed to be included as input data of building management for energy efficiency. Then, Section 7.3 reviews the main related works which propose partial solutions to the problem addressed in this chapter. Section 7.4 presents a general architecture proposal for management systems of smart buildings, which is modeled in three layers with different functionalities. Section 7.5 describes our proposal for an energy efficiency building management system. This proposal tackles three different subproblems, each one of these is introduced here. Section 7.6 summarizes the experiments carried out to evaluate and validate the different proposals and mechanisms developed in this work. Finally, Section 7.7 gives some conclusions and an outlook of future work.

7.2 Addressing Energy Efficiency in Smart Buildings

Optimizing energy efficiency in buildings is an integrated task that comprises the whole lifecycle of the building. For buildings to have an impact at city level in terms of energy efficiency, different challenges have been identified in the building value chain (from design to end-of-life of buildings)[2], which can be summarized as follows:

1. *Design*. The design of buildings should be integrated, holistic and multi target.

[2]http://www.ectp.org/

2. *Structure*. The structure of buildings should provide features such as safety, sustainability, adaptability and affordability.
3. *Building envelope*. This should ensure efficient energy and environmental performance. Prefabrication is a crucial step to guarantee energy performance. Multifunctional and adaptive components, surfaces and finishes to create added energy functionality, and durability should all be built in.
4. *Energy equipment and systems*. Advanced heating/cooling and domestic hot water solutions, including renewable energy sources, should focus on sustainable generation as well as on heat recovery. Among these systems, thermal storage (both heat and cold) is recognized as a major breakthrough in building design. Distributed/decentralised energy generation should address the key requirement of finding smart solutions for grid-system interactions on a large scale. ICT smart networks will form a key component in such solutions. In [10], for instance, the authors study the communication requirements for smart grids and describe the most suitable communication protocols, wired and wireless, with special attention to the latest proposals in this field.
5. *Construction processes*. These should consider ICT-aided construction, improving the energy performance delivered, and automated construction tools.
6. *Performance monitoring and management*. This should ensure interoperability among the different subsystems of the building, including smart energy management systems that provide flexible actions to reduce the gap between predicted and actual energy building performance, occupancy modeling, the fast and reproducible assessment of designed or actual performance, and continuous monitoring and control during service life. Finally, knowledge sharing must be considered by means of open data standards that allow collaboration among stakeholders and interoperability among systems.
7. *End of life*. This should include decision-support concerning possible renovation or the construction of a new building and associated systems.

During these phases it is necessary to continuously re-engineer the indexes that measure energy efficiency to adapt the energy management system to the building's conditions. Hereinafter, we refer only to electrical energy consumption since other kinds of energy such as fuel, gas or water are beyond the scope of this work. Taking as reference the energy performance model for buildings proposed by the CEN Standard EN15251 [11], it proposes criteria for dimensioning the energy management of buildings, while indoor environmental requirements are maintained. According to this standard, there are static and dynamic conditions that affect the energy consumption of buildings. Given that each building has a different static model according to its design, we try to provide a solution for energy efficiency focusing on analyzing how dynamic conditions affect the energy consumed in buildings. Thus, we propose an initiative for the challenges involved in the living stage buildings: *Performance monitoring and management* mentioned in the above list. In this stage, we need to identify the main drivers of energy use in buildings. After monitoring these parameters and analysing the associated energy consumed, we can model their impact on energy consumption, and then, propose control strategies to save energy. The main idea of this approach is to provide anticipated responses to ensure energy efficiency in buildings.

Bearing in mind all these concerns, we enumerate below the stages [12] that must be carried out to achieve efficiency building energy management:

1. **Monitoring**. During the monitoring phase, information from heterogeneous sources is collected and analysed before concrete actions are proposed to minimize energy consumption, bearing in mind the specific context of a given building. Since buildings with different functionalities have different energy use profiles, it is necessary to carry out an initial characterization of the main contributors to their energy use. For instance, in residential buildings the energy consumed is mainly due to the indoor services provided to their occupants (associated to comfort), whereas in industrial buildings energy consumption is associated mostly with the operation of industrial machinery and infrastructures dedicated to production processes. Considering this, and taking into account the models for predicting the comfort response of buildings occupants given by the ASHRAE [13], we describe below the main parameters that must be monitored and analysed before implementing optimum building energy management systems. In this way, from this set of parameters affecting energy consumption in buildings, we can extract the input data to be taken into account.

(a) Electrical devices always connected to the electrical network. In buildings, it is necessary to characterize the minimum value of energy consumption due to electrical devices that are always connected to the electrical network, since they represent a constant contribution to the total energy consumption of the building. For this, it is necessary to monitor over a period of time the energy consumed in the building when there is no other contributor to the total energy being consumed. This value will be included as an input to the final system responsible for estimating the daily electrical consumption of the building.

(b) Electrical devices occasionally connected. Depending on the kind of building under analysis, different electrical devices may be used with different purposes, such as increase of productivity and comfort. On the other hand, the operation of such devices could be independent of the participation and behaviour of the occupants; for example, in the context of a factory or an office where there are timetables and rules. Whatever the case, recognition of the operation pattern of devices must be included in the final system responsible for estimating the daily electrical consumption of the building. To obtain these patterns it is necessary to monitor previously the associated energy consumption of every device or appliance. To monitor each component separately in the total power consumption in a household or an industrial site over time, cost effective and readily available solutions include Non-Intrusive Load Monitoring (NILM) techniques [14].

(c) Occupants' behaviour. Energy consumption of buildings due to the behaviour of their occupants is one of the most critical points in every building energy management system. This is mainly because occupant behaviour is difficult to characterize and control due to its uncertain dynamic. First of all, it is necessary to have solved the occupants' localization before behaviour models associated to them can be provided. Depending on the building context, the impact of occupants behaviour on total energy consumption is different. For example, in residential buildings the impact of the behaviour in the energy consumed is one of the biggest, followed by environmental conditions. However, in buildings with productive goals, the electricity consumed by the appliances and devices working for such goals is usually the main contributor to the total energy consumed in the building. Therefore, it is necessary to monitor and analyse this issue to be able to provide behaviour patterns that will be included in the final estimation of the daily energy consumption of the building. To do so, different techniques, like crowd sensing, can be used to extract a palette of underlying behavioural patterns. In that sense, occupants' behaviour can be characterized for features such as:

- Occupants localization data.
- Activity level of occupants.
- Comfort preferences of occupants.

(d) Environmental conditions. Parameters like temperature, humidity, pressure, natural lighting, etc. have a direct impact on the energy consumption of buildings. Nevertheless, depending on the specific context of the building and its requirements, this impact will differ and be greatest in the case of indoor comfort services (like thermal and visual comfort). Therefore, forecasts of the environmental condition should be also considered as input for the final energy consumption estimation of the building.

(e) Information about the energy generated in the building. Sometimes, alternative energy sources can be used to balance the energy consumption of the building. Information about the amount of daily energy generated and its associated contextual features can be used to estimate the total energy generated in the future. This information allows us to design optimal energy distribution or/and strategies of consumption to ensure the energy-efficient performance of the building.

(f) Information about total energy consumption. Knowing the real value of the energy consumed hourly or even daily permits the performance and accuracy of the building energy management program to be evaluated, and make it possible to identify and adjust the system in case of any deviation between the consumption predicted and the real value. In addition, providing occupants with this information is crucial to make them aware of the energy that they are using at any time, and encourage them to make their behaviour more responsible.

In this work we focus on residential buildings, where both comfort and energy efficiency is required. As regards the comfort provided in buildings, we focus on thermal and visual comfort.

2. **Information Management**. An intelligent management system must provide proper adaptation countermeasures for both automated devices and users with the aim of providing the most important services in buildings (comfort) and satisfying energy efficiency requirements. Therefore, energy savings needs to be addressed by establishing a trade-off between the quality of services provided in buildings and the energy resources required for the same, as well as the associated cost.

3. **Automation**. Automation systems in buildings take inputs from the sensors installed in corridors and rooms (light, temperature, humidity, etc.), and use these data to control certain subsystems such as HVAC, lighting or security. These and more extended services can be offered intelligently to save energy, taking into account environmental parameters and the location of occupants. Therefore, automation systems are essential to answer the needs for monitoring and controlling energy efficiency requirements [15]. At this respect, the 1888–2011 IEEE Standard for Ubiquitous Green Community

Control Network Protocol [16] describes remote control architecture of digital community, intelligent building groups, and digital metropolitan networks; specifies interactive data format between devices and systems; and gives a standardized generalization of equipment, data communication interface, and interactive message in this digital community network.

4. **Feedback and User Involvement.** Feedback on consumption is necessary for energy savings and should be used as a learning tool. Analysis of smart metering, which provides real-time feedback on domestic energy consumption, shows that energy monitoring technologies can help reduce energy consumption by 5% to 15% [5]. As can be deduced, a set of subsystems should be able to provide consumption information in an effective way. These subsystems are:

- Electric lighting.
- Boilers.
- Heating/cooling systems.
- Electrical panels.

On the other hand, to date, information in real-time about building energy consumption has been largely invisible to millions of users, who had to settle with traditional energy bills. In this, there is a huge opportunity to improve the offer of cost-effective, user-friendly, healthy and safe products for smart buildings, which increase the awareness of users (mainly concerning the energy they consume), and permit them to bean input of the underlying processes of the system. This would allow the collection of an unprecedented amount of data related to users' interactions and their associated contextual details (e.g. identity, location and activity) by considering the active involvement of users along with opportunistic sensing. Then, an appropriated processing of that user-related data will enable the development of even more customized services.

Taking into account all the aspects identified as relevant for their impact in energy consumption of buildings, we review how related works from the literature tackle them. In this way, we can extract the main limitations and constraints of these works, and suggest proposals to address them.

7.3 Related Work

A complete review of previous solutions from the literature was carried out during the development period of the present chapter. We tried to find ways that would enable us to propose holistic solutions to building energy management problems, which should address there relevant aspects mentioned previously, i.e. a complete monitoring phase, the efficient management of information, using automation systems and involving occupants during the system operation. Nevertheless, different proposals were found for different goals, but none was integrated all the aspects. This was the first constraint identified among previous solutions. Consequently, we decided to review the main related work tackling each one of these aspects separately.

As regards the monitoring aspect, initial solutions to energy efficiency in buildings were mainly focused on non-deterministic models based on simulations. A number of simulation tools are available with varying capabilities. In [17] and [18] a comprehensive comparison of existing simulation tools is provided. Among these tools are ESP-r [19] and Energy Plus [20]. However, this type of approach relies on very complex predictive models based on static perceptions of the environment. For example, a multi-criteria decision model to evaluate the whole lifecycle of a building is presented in [21]. The authors tackle the problem from a multi-objective optimization viewpoint, and conclude that finding an optimal solution is unrealistic, and that only an approximation is feasible.

With the incessant progress made in the field of ICT and sensor networks, new applications to improving energy efficiency are constantly emerging. For instance, in office spaces, timers and motion sensors provide a useful tool to detect and respond to occupants, while providing them with feedback information to encourage behavioural changes. The solutions based on these approaches are aimed at providing models based on real sensor data and contextual information. Intelligent monitoring systems, such as automated lighting systems, have limitations such as those identified in [22], in which the time delay between the response of these automated systems and the actions performed can reduce any energy saving, whilst an excessively fast response can produce inefficient actions. These monitoring systems, while contributing towards energy efficiency, require significant investment in an intelligent infrastructure that combines sensors and actuators to control and modify the overall energy consumption. The cost and difficulty involved in deploying such networks often constrain their viability. Clearly, an infrastructure-less system that uses existing technologies would provide a cheaper alternative to building energy management systems. On the other hand, building energy management must bare with the inaccuracy of sensors, the lack of adequate models for many processes and the non-deterministic aspects of human behaviour.

In this sense, there is an important research area that proposes techniques of artificial intelligence as a way of providing intelligent building management systems. Rather than solving the above drawbacks. This approach involves models based on a combination of real data and predictive patterns that represent the evolution of the parameters affecting the energy consumption of buildings. An example of such an approach is [23], in which the authors propose an intelligent system able to manage the main comfort services provided in the context of a smart building, i.e. HVAC and lighting, while user preferences concerning comfort conditions are established according to the occupants' locations. Nevertheless, the authors only propose the inputs of temperature and lighting in order to make decisions, while many more factors are really involved in energy consumption and should be included to provide an optimal and more complete solution to the problem of energy efficiency in buildings. Furthermore, no automation platform is proposed as part of the solution.

Regarding building automation systems, many works extend the domotics field which was originally used only for residential buildings. A relevant example is the proposal given in [24], where the authors describe an automation system for smart homes based on a sensor network. However, the system proposed lacks automation flexibility, since each node of the network offers limited I/O capabilities through digital lines, i.e. there is no friendly local interface for users, and most importantly, integration with energy efficiency capabilities is weak. The work presented in [25] is based on a sensor network to cope with the building automation problem for control and monitoring purposes. It provides the means for open standard manufacturer-independent communication between different sensors and actuators, and appliances can interact with each other with defined messages and functions. Nevertheless, the authors do not propose a control application to improve energy efficiency, security or living conditions in buildings.

The number of works concerning energy efficiency management in buildings using automation platforms is more limited. In [26], for instance, a reference implementation of an energy consumption framework is provided, but it only analyses the efficiency of ventilation system. In [27] the deployment of a common client/server architecture focused on monitoring energy consumption is described, but without performing any control action. A similar proposal is given in [28], with the main difference that it is less focused on efficiency indexes, and more on cheap practical devices to cope with a broad pilot deployment to collect the feedback from users and address future improvements for the system.

Regarding commercial solutions for the efficient management of building infrastructures, there are proposals such as those given by the manufacturer

Johnson Controls[3], a company that provides products, services and solutions that help increase energy efficiency and reduce the operation costs of its clients' buildings. Another well-known manufacturer is Siemens[4], who offer a technical infrastructure for building automation and energy efficiency in the form of market-specific solutions in buildings and public places. The main differences between these commercial solutions and our proposal for automation and energy efficiency management in smart buildings are those related with the open and transparent character of our proposal, as well as its capability to gather data from a large number of heterogeneous sources.

As regards user involvement, this can be done by means of their implicit or explicit feedback. When implicit feedback is considered, an important line of research focuses on the crowdsensing paradigm [9]. In brief, this paradigm intends to uncover meaningful behavioural patterns by automatically collecting the digital breadcrumbs of the different sensors that users' personal devices are equipped with. At the same time, a novel course of action has paid attention to social networks as a novel datasource to extend the collection implicit user feedback [29]. Despite its inherent uncertainty, several works are already able to extract meaningful behavioural patterns by mainly using social-network feeds [30, 31]. As for explicit user's feedback, the crowdsourcing paradigm centers on providing tools to allow the management of the information explicitly requested to sets of target users [32, 33]. In a smart building context, crowdsensing or crowdsourcing paradigms have been mainly used to flow management in indoor areas [34]. Last but not least, in the building energy management field, some proposals have involved uses in saving energy in buildings [5, 6]. However, few works have been addressed this aspect. It is important to note that energy usage feedback in building energy management systems needs to be provided to users frequently and over a long time, offering an appliance-specific breakdown, while presented in a clear and appealing way using computerized and interactive tools.

Concerning the fact that users have little awareness of the energy wastage associated with their energy consumption behaviours is due partly to the fact that most people do not know what the optimum comfort conditions are according to environmental features and their needs. It is clear that, while each person has his/her own comfort preferences and these preferences are strongly conditioned by subjective concerns, there are a minimal and a maximum set of

[3]http://www.johnsoncontrols.co.uk/content/gb/en/products/building_ efficiency.html

[4]http://www.buildingtechnologies.siemens.com/bt/global/en/energy-efficiency/Pages/Energy-efficiency.aspx

comfort conditions recognized as common to everyone to ensure the quality of life [35]. Therefore, the confidence and respect that users give to the intelligent services that are offered to them in terms of comfort and energy efficiency concerns in smart buildings, are crucial constraints in this type of system. Nevertheless, thanks to pervasive computing practices, the integration and development of systems based on IoT support and encourage the cooperation between humans and devices in terms of:

- Facilitating communication between things and people, and between things, by means of a collective network intelligence context.
- People's ability to exploit the benefits of this communication through their increasing familiarity with ICT.
- A vision where, in certain respects, people and things are homogeneous agents endowed with fixed computational tools.

Smart buildings should prevent users from having to perform routine and tedious tasks to achieve comfort, security, and effective energy management. Sensors and actuators distributed in buildings can make user life more comfortable; for example: i) room heating can be adapted to user preferences and to the weather; ii) room lighting can change according to the daylight; iii) domestic incidents can be avoided with appropriate monitoring and alarm systems; and, iv) energy can be saved by automatically switching off electrical equipment when not needed, or regulating their operating power according to user needs, thus avoiding any energy overuse. In this sense, IoT is a key enabler of smart services to satisfy the needs of individual users, who apart from being users of the system, can also be seen as sensors in the same way as temperature, thermal, humidity and presence sensors deployed in the building.

As can be noted, most of the approaches proposed to date only provide partial solutions to the overall problem of energy efficiency in buildings, where, although different factors are involved holistically, until now they have been addressed separately or even neglected by previous proposals. This division is frequently due to the uncertainty and lack of data and inputs in the management processes, so that analysis of how energy in buildings is consumed is incomplete. In other words, a more integral vision is required to provide accurate models of the energy consumed in buildings [7]. In this sense, no solutions have been proposed tackling the full integration of information related with all relevant aspects directly involved in the energy consumption of buildings (which are described in Section 7.2). For example, there are not previous solutions that fully integrate information about the occupants of buildings, despite of the fact that human behaviour has been recognized as one of the most important aspects affecting energy consumption in buildings. Information about the identities of occupants, their locations and activities, their comfort preferences, their levels of awareness with the problem of the high energy consumption of buildings, their participation to get energy saving, etc. must be included, jointly to other relevant information, in any building energy management system. In this chapter, we present our own smart system proposal, which is a holistic and flexible solution based on collecting and analysing information of both the building context and its occupants, and propose concrete actions which could be applied in the management of any controllable infrastructure of buildings to ensure their energy efficient performance. Our proposal of solution considers occupants as a key piece of our management system, and we demonstrate the benefits of following this approach in term of the energy saving achieved in various buildings used as reference.

7.4 A Proposal of General Architecture for Management Systems of Smart Buildings

The architecture of our proposal for smart building is modelled in layers which are generic enough to cover the requirements of different smart environments of cities, such as intelligent transport systems, security, health assistance or, as is the case analysed in this chapter, smart buildings. This architecture promotes high-level interoperability at the communication, information and services layers. The layers of such architecture are depicted in Figure 7.1, and are detailed below.

7.4.1 Data Collection Layer

Looking at the lower part of Figure 7.1, input data are acquired from a plethora of sensor and network technologies such as the Web, local and remote databases, wireless sensor networks, mobile devices, etc., all of them forming an IoT ecosystem. In this sense, and considering the instance of this architecture for the building management system proposed in this chapter, it gathers information from sensors and actuators deployed in the building. As for static sensors and actuators can be self-configured and controlled remotely through the Internet, enabling a variety of monitoring and control applications. Concerning mobile sensors, mechanisms to pro-actively or passively collect their reported data is also included in this layer. Given the heterogeneity of data sources and the necessity of seamless integration of devices and networks,

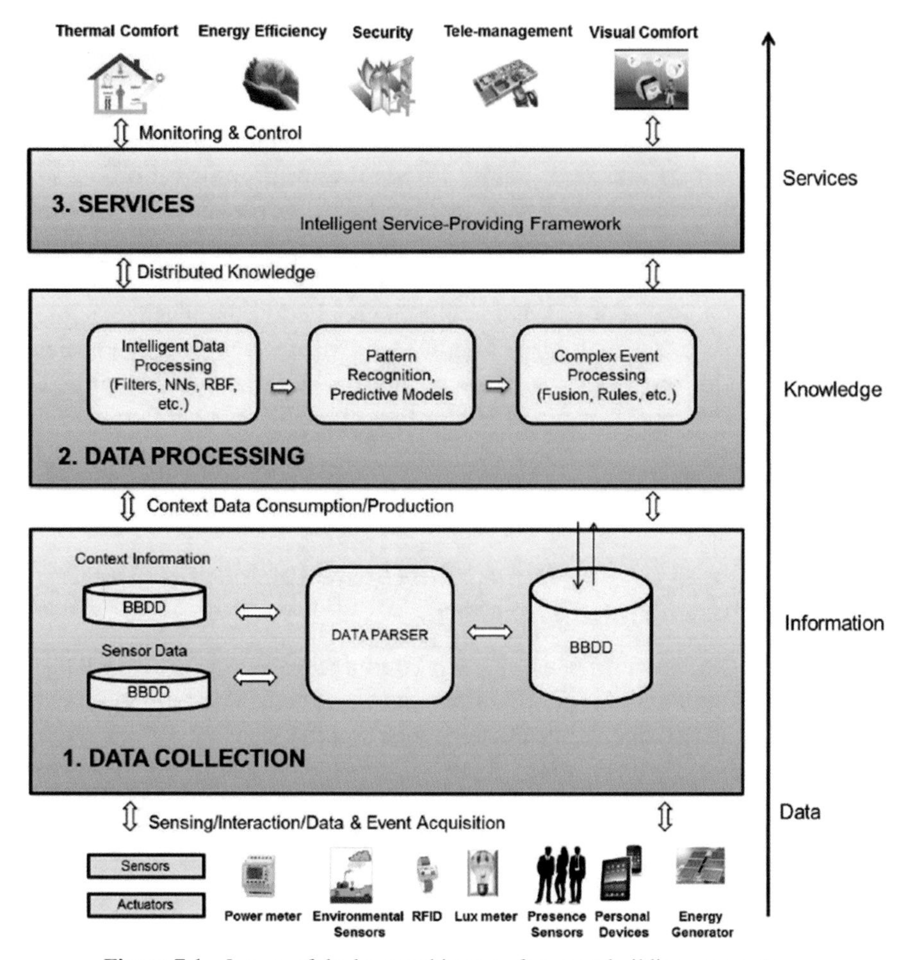

Figure 7.1 Layers of the base architecture for smart buildings ecosystem.

a common language structure to represent data is needed to deal with this issue. Therefore, the transformation of the collected data from the different data sources into a common language representation is performed in this stage.

7.4.2 Data Processing Layer

The data processing layer is responsible for processing the information collected and making decisions according to the final application context. A set of information processing techniques is applied to extract, contextualize, fuse and represent information for the transformation of massive input data into useful knowledge, which can be distributed later towards the services layer. Different algorithms can be applied for the intelligent data processing and decision making processes, depending on the final desired operation of the system (i.e. the services addressed). Considering the target application of smart buildings, data processing techniques for covering, among others, security, tele-assistance, energy efficiency, comfort and remote control services should be implemented in this layer. And following a user-centric perspective for services provided, intelligent decisions are made through behaviour-based techniques to determine appropriate control actions, such as appliances and lights, power energy management, air conditioning adjustment, etc.

7.4.3 Services Layer

Finally, the specific features for providing services, which are abstracted from the final service implementation, can be found in the upper layer of the proposed architecture (see Figure 7.1). Our approach offers a framework with transparent access to the underlying functionalities to facilitate the development of different types of final application. This generic proposal of architecture for smart buildings has been instantiated in the system known as City explorer. City explorer, which was developed at the University of Murcia, integrates an automation platform which is divided into an indoor part, and all the connections with external elements for remote access, technical tele-assistance, security and energy efficiency/comfort providing services in buildings. Figure 7.2 shows a schema of City explorer offering ubiquitous services in the smart buildings field. The main components of City explorer were presented in details in [36, 37]. The work developed in this chapter is based on using City explorer as platform of experimentation and validation of our proposal of building management to achieve energy efficiency. For this, we have instantiated each generic layer of the architecture shown in Figure 7.1, with the goal of offering a solution to energy efficiency in smart buildings.

7.5 IoT-based Information Management System for Energy Efficiency in Smart Buildings

As mentioned before, our proposal of IBMS uses the City explorer platform applied to achieve energy efficiency in buildings. Our proposed system has the capability, among others, to adapt the behaviour of automated devices

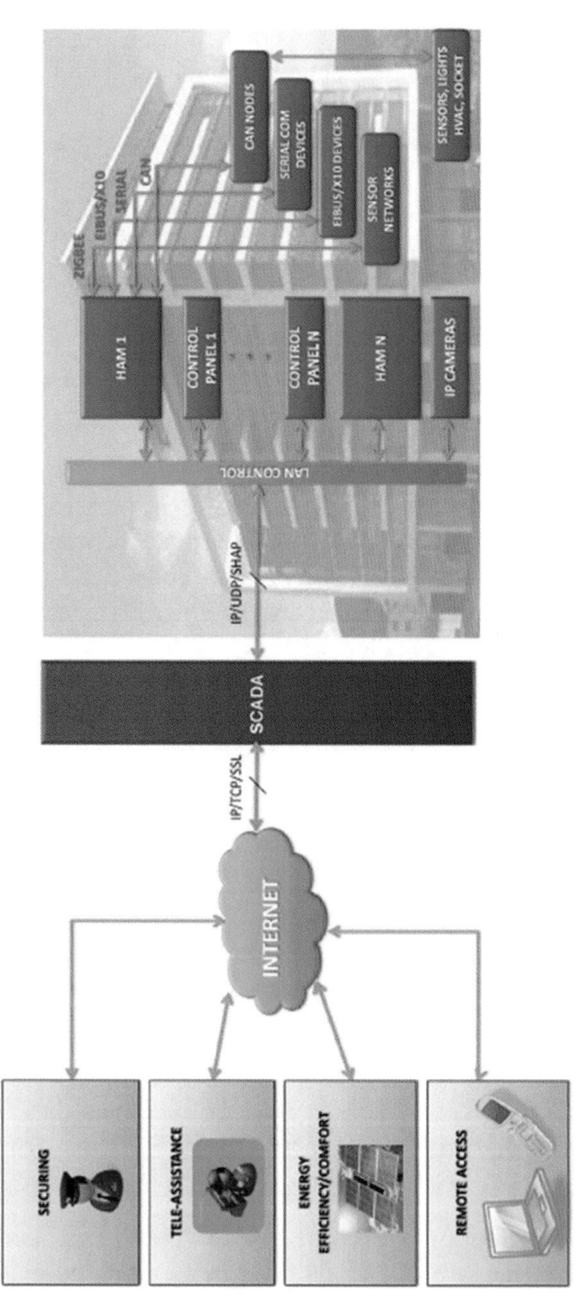

Figure 7.2 City explorer applied to smart buildings.

deployed in the building in order to meet energy consumption restrictions, while maintaining comfort conditions at the occupants' desired levels.

More specifically, the goals of our intelligent management system are the following:

- High comfort level: learn the comfort zone from users' preferences, guarantee a high comfort level (thermal, air quality and illumination) and a good dynamic performance.
- Energy savings: combine the control of comfort conditions with an energy saving strategy.
- Air quality control: provide CO2-based demand-controlled ventilation systems.

Satisfying the above control requirements implies controlling the following actuators:

- Shading systems to control incoming solar radiation and natural light as well as to reduce glare.
- Window opening for natural ventilation or mechanical ventilation systems to regulate natural airflow and indoor air changes, thus affecting thermal comfort and indoor air quality.
- Heating/cooling (HVAC) systems.

As a starting point, we focus only on the management of lights and HVAC subsystems, since they represent the highest energy consumption at building level. User interactions have a direct effect on the whole system performance, because the occupants can take control of their own environment at any time.

Thus, the combined control of the system requires optimal operation of every subsystem (lighting, HVAC, etc.), on the assumption that each operates normally in order to avoid conflicts arising between users' preferences and the simultaneous operations of such subsystems. Figure 7.3 shows a schema of the different subsystems comprising the intelligent management system integrated in City explorer, where the outputs of the system are forwarded to the actuators deployed in the building.

As can be seen in Figure 7.3, the first task to solve is related with user identification and localization, and the second problem is related with the issues of comfort and energy efficiency in the management of the building. In the following subsections we describe the different issues involved and which were solved during this work, and represent our proposal of building energy management system for energy efficiency.

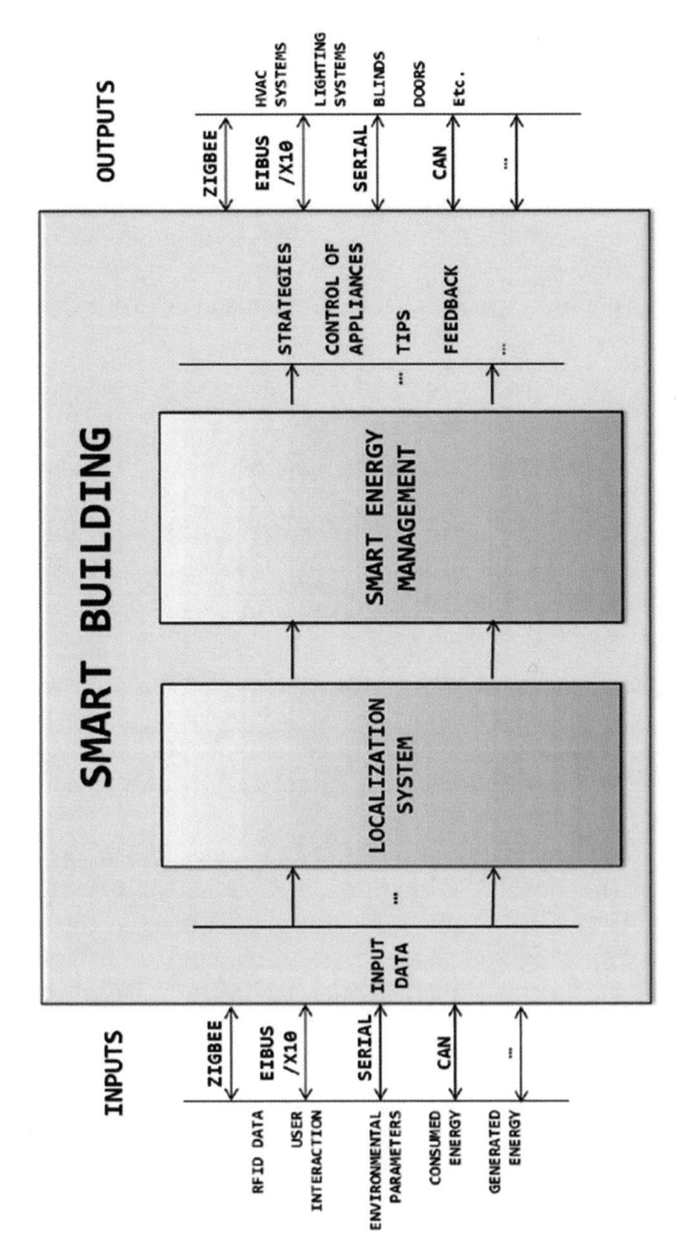

Figure 7.3 Schema of the modules composing the management system in charge of the building comfort and energy efficiency.

7.5.1 Indoor Localization Problem

In a smart building, embedded sensors measure and record user activities, making it possible to predict their future behaviour, prepare everything one step ahead according to the individual user's preferences or needs, and provide the most convenient energy efficient services. These services need to operate by acquiring contextual information both from users and the environment. Therefore, to make buildings smart and to be able to offer users customized services, it is indispensable to previously solve the implicit indoor localization problem. Furthermore, user identities need to be taken into account so that the intelligent system can learn and manage devices according to their behaviour and/or preferences. We obviously need to solve user identification in smart buildings to provide customized comfort services committed to energy efficiency, but while user privacy must also be respected because occupants care about their private and social activities, and want full control of how their personal location information and history are used. Hence, there is a need to rely on non-intrusive, ubiquitous and cheap sensors to minimise infrastructure deployment and prevent user dissatisfaction. Indeed, some sensors cannot be installed in buildings; for instance, in Spain video cameras cannot be legally used in offices. Problems like this make some localization systems unsuitable for use in smart buildings.

In the scenario addressed in this work, the whole area of a smart building is divided into locations (rooms, open areas, corridors, etc.) with different comfort conditions in each one; for instance, optimum lighting conditions in a corridor are different from those required in an office; or the optimum level of air conditioning in an individual bedroom is different from that required in a very crowded dining room. Furthermore, in each of these areas (an individual bedroom, a dining room, an office, etc.), it is necessary to carry out a further division depending on the service area of each comfort appliance deployed. Therefore, our indoor localization system must be able to locate a user in terms of regions, which correspond to the service areas of the appliances or devices involved in her/his comfort condition. Recent years have seen great progress in indoor localization systems, but there are still some weaknesses in terms of the accuracy of location data, the time required for calibration processes, poor robustness, or high installation and equipment costs [38]. Furthermore, when user identification is needed, most of the systems proposed present difficulties concerning complexity, computational load and inaccurate results. Since the indoor localization problem does not have obvious solutions, we review relevant solutions from the literature and identify the technological options most suitable in light of our problem. Accuracy is usually the most important requirement for positioning systems. In the location problem involved in energy efficiency of buildings, we conclude that the accuracy required for our localization system depends on the service areas of the appliances and devices involved in the comfort and energy balance of the building.

In Figure 7.4 a rough outline of some positioning systems is presented, with their accuracy ranges achieved until now according to the literature. Since each localization technology has its particular advantages and disadvantages, we suggest that by combining several complementary technologies and applying data fusion techniques, it is possible to improve the overall system performance and provide a more reliable indoor localization system, since more specific inferences can be achieved than when using a single kind of data sensor. Therefore, after analysing Figure 7.4, we choose a hybrid solution based on RF and non-RF technologies. Our technological solution to cover the localization needs (i.e. those required by smart buildings to provide occupants with customized comfort services) is based on a single active RFID system and several Infra-Red (IR) transmitters. In Figure 7.5 we can observe the data exchange carried out among the different technological devices that compose our localization system.

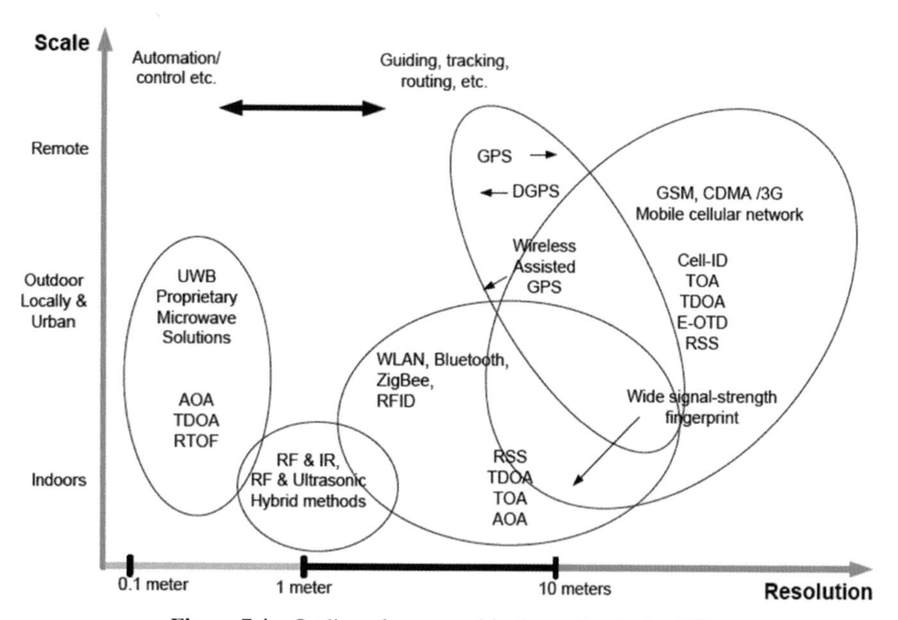

Figure 7.4 Outline of some positioning technologies [38].

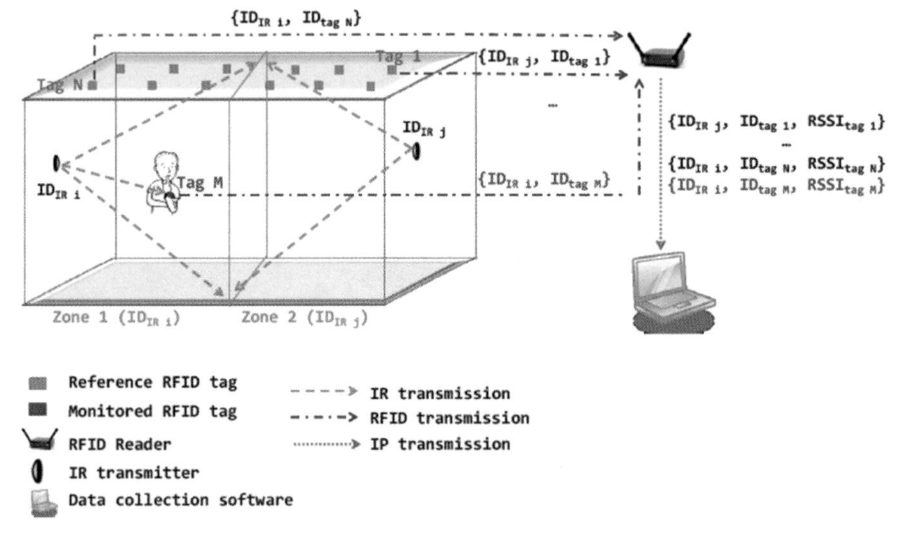

Figure 7.5 Localization scenario.

The final mechanism implemented for indoor localization is shown in Figure 7.6. In this figure, we can see that the first phase of our localization mechanism is the space division through the installation of IR devices in the walls of the building area where localization wants to be solved. Therefore, for each space division, there is an IR identifier value (IDir) associated to this region. For each one of these region, we implement a regression method based on Radial Basis Functions (RBF) networks. The RBF estimates user positions given different RFID tags situated in the roof. This RFID-based information coming from the different building's occupants conforms a data stream that could be also processed by means of a crowdsensing approach so as to track the flow of people within a building. In that sense, several proposals already exist that intends to reconstruct the behaviour of people by using the type of discrete locations [39].

In our localization mechanism, after the position estimation using the RBF network, a Particle Filter (PF) is applied as a monitoring technique, which takes into account previous user position data for estimating future states according to the current system model. In the PF, we modify particle weights according to the distances to the measurements during the correction stage, as the following equation shows:

$$w(\overrightarrow{x}_t) = w(\overrightarrow{x}_{t-1}) \cdot \frac{p(\overrightarrow{y}_t|\overrightarrow{x}_t) \cdot p(\overrightarrow{x}_t|\overrightarrow{x}_{t-1})}{q(\overrightarrow{x}_t|\overrightarrow{x}_{t-1}, \overrightarrow{y}_t)} \tag{7.1}$$

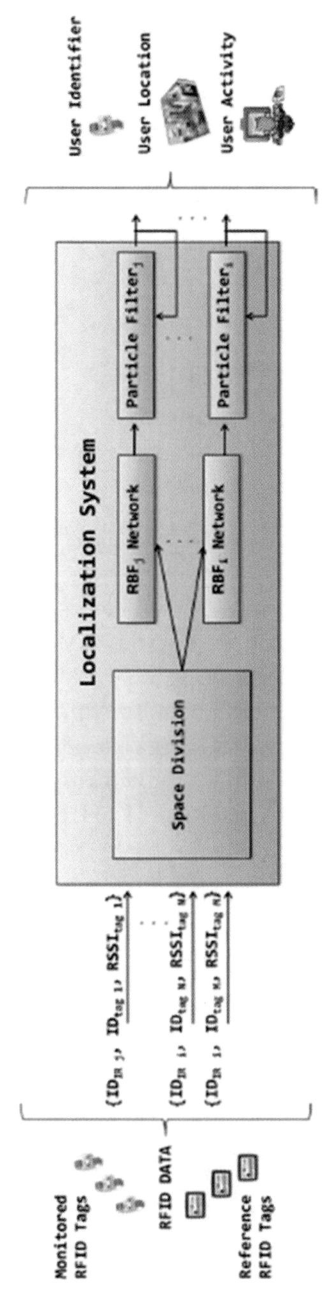

Figure 7.6 Data processing for location estimation.

where $w(\overrightarrow{x}_t)$ weights of the set of particles at instant t; $p(\overrightarrow{y}_t|\overrightarrow{x}_t)$ and $p(\overrightarrow{x}_t|\overrightarrow{x}_{t-1})$ gives the probabilistic behaviour of the output and the state model of the system respectively, and $q(\overrightarrow{x}_t|\overrightarrow{x}_{t-1}, \overrightarrow{y}_t)$ is the approximation of the expectedly function.

Algorithm 7.1 provides a summarized version of the general definition of PF. The PF used in this work is slightly different from its generic definition. The main difference of our filtering algorithm is in the correction stage, which applies the resample using the Sequential Importance Sampling (SIS) algorithm [40] (step 13 of Algorithm 7.1). During this step, information about the specific IR region at a given instant of time is also used to benefit those particles which fall inside this area. Therefore, before applying Equation (7.1), we filter according to the condition given by Equation (7.2):

$$\{\text{If}: y_t \in \Omega^j \Rightarrow w(x_t^i) = 0 \,\forall\, x_t^i \notin \Omega^j\}, \tag{7.2}$$

where Ω^j represents the coverage area of the IR transmitter with identifier j, and y_t and $w(x_t^i)$ denote, respectively, the measured parameter and the weight of the set of particles *i* at the instant of time *t*. The main advantage of this constraint is the faster convergence of the filter, because extra information is available to carry out the correction stage.

Algorithm 7.1 Generic PF

Require: $\{x_{t-1}^i, w_{t-1}^i\}_{i=1}^{N_s}, y_t$
Ensure: $\{x_t^i, w_t^i\}_{i=1}^{N_s}$
1: Given a particle number N_s
2: Given a threshold N_T value for resampling
3: **for** *i* = 1 **to** N_s **do**
4: Draw $x_t^i \sim q(x_t|x_{t-1}^i, y_t)$
5: Assign the particle a weight w_t^i
6: **end for**
7: Calculate total weight: *t* = SUM$[\{w_t^i\}_{i=1}^{N_s}]$
8: **for** *i* = 1 **to** N_s **do**
9: Normalize: $w_t^i = t^{-1} \cdot w_t^i$
10: **end for**
11: calculate $\widehat{N_{eff}} = \frac{1}{\sum_{i=1}^{N_s}(w_t^i)^2}$
12: **if** $\widehat{N_{eff}} \leq N_T$ **then**
13: Correction stage.
14: **end if**

7.5.2 Building Energy Consumption Prediction

The energy performance model of our BMS is based on the CEN Standard N15251 [41]. This standard proposes the criteria of design for any building energy management system. It establishes and defines the main input parameters for estimating building energy requirements and evaluating the indoor environment conditions. The inputs considered to solve our problem are the data coming from the RFID cards of users, the user interaction with the system through the control panels or the web access, environmental parameters coming from temperature, humidity and lighting sensors installed in outdoor and indoor spaces, the consumption energy sensed by the energy meters installed in the building, and the generated energy sensed by the energy meters installed in the solar panels deployed in our testbed. After collecting the data, it is mandatory to continue with their cleaning, preprocessing, visualization and correlation study in order to find determining features, which can be used to generate optimal energy consumption models of buildings (management layer of the architecture presented in Section 7.4). Over the input set, we perform the standardization and reduction of data dimensionality using Principal Components Analysis (PCA) [42], identifying the directions in which the observations of each parameter mostly vary.

Regarding the Artificial Intelligence (AI) techniques that have been already applied successfully to generate energy consumption models of buildings in different scenarios (as such we mentioned in the management layer of the architecture presented in Section 7.4), we propose to evaluate the performance of Multilayer Perceptron (MLP), Bayesian Regularized Neural Network (BRNN) [43], SVM [44] and Gaussian Processes with RBF Kernel [45]. They were selected because of the good performance that all of them have already provided when they are applied to building modelling. All these regression techniques are implemented following a model-free approach, which is based on selecting – for a specific building – the optimal input set and technique, i.e. such input set and technique that provides the most accurate predictive results in a test data set. In order to implement this free-model approach, we use the R [46] package named CARET [47] to train the energy consumption predictive algorithms, looking for the optimal configuration of their hyper-parameters.

The selected metric to evaluate the models generated for each technique using test sets is the well-known RMSE (Root-Mean-Square Error), which formulation appears in Equation (7.3).

$$RMSE = \sqrt{\frac{1}{n}\sum_{i=1}^{n}(y_i - \hat{y}_i)^2} \tag{7.3}$$

This metric shows the error by means of the quantity of KWh that we deviate when predicting, but in order to get a better understanding of the uncertainty of the model, we also show its coefficient of variation (CVRMSE). This coefficient is the RMSE divided by the mean of the output variable (energy consumption) for the test set (Equation (7.4)), giving us a percentage of error adjusted to the data, not just a number in general terms.

$$CVRMSE = \frac{RMSE}{\overline{y}} \tag{7.4}$$

7.5.3 Optimization Problem

Once the building energy consumption is modeled we focus on the optimization of its use trying to keep comfort conditions. As starting point, we establish the comfort extremes considering location type, user activity and date [48]. Understanding the building thermal and energetic profiles allows us to quantify the effects of particular heating-cooling set point decisions. To derive a heating or cooling schedule, it is necessary to formulate the target outcome. In our buildings, it is possible to:

1. Optimize the indoor temperature during occupation, i.e. minimize the building temperature deviations from a target temperature.
2. Minimize daily energy consumption, or
3. Optimize a weighted mixture of the criteria, a so-called multi-objective optimization.

The definition of building temperature deviation influences the results strongly: taking the minimum building temperature will result in higher set point choices and higher energy use than using e.g. the average of building temperatures. Constraints on maximum acceptable deviation from target comfort levels or an energy budget can be taken into account to ensure required performance. For our optimization problem, we apply a genetic optimization implemented in R (using the "genalg" package [49]) to our predictive building models to derive schedules for heating/cooling setpoints.

7.5.4 User Involvement in the System Operation

Following this approach to provide human-centric services in the context of smart buildings, users can be seen as both the final deciders of actions, and system co-designers in terms of feedback that conditions future rules and contributions to the software issuing these rules. In this sense, in our energy building management system we consider the data provided directly by users through their interactions when they change the comfort conditions provided automatically by the system and, consequently, the system learns and autoadjusts according to such changes and to the control comfort/energy strategies defined by users using the graphic editor of City explorer. Furthermore, with the aim of offering users information about any unsuitable design or setting of the system, as well as to help them easily understand the link between their everyday actions and environmental impact, City explorer is able to notify them about such matters (i.e. acting as a learning tool). On the other hand, when the system detects disconnections and/or failures in the system, it sends alerts by email/messages to notify users to check these issues. All these features, which are included in our management system, contribute to user behaviour changes and increase their awareness as time passes, or detect unnecessary stand-by consumption of the controllable subsystems of the building.

Finally, to understand the background of energy behaviour of users involved in our experiments and to be able to form an initial context pattern for the usability of the system under different constraints, we carried out a follow-up study based on the feedback that users provide to City explorer through the SCADA-web and the control panels installed in the smart building. Another reason to carry out this study was the identified lack of research in the building energy management area, where large-scale deployment needs to be accompanied by a body of study on user behaviour, motivation and preferences. The same was pointed out by [6]. In Figure 7.7 is shown the schema of our final building energy management solution.

7.6 Evaluation and Results

7.6.1 Scenario of Experimentation

The reference building where our BMS for energy efficiency is deployed is the Technology Transfer Centre (TTC) of the UMU[5]. Every room of this building is automated through a Home Automation Module (HAM) unit of the City explorer platform. It permits us to consider a granularity at room level to carry out the experiments.

[5]www.um.es/otri/?opc=cttfuentealamo

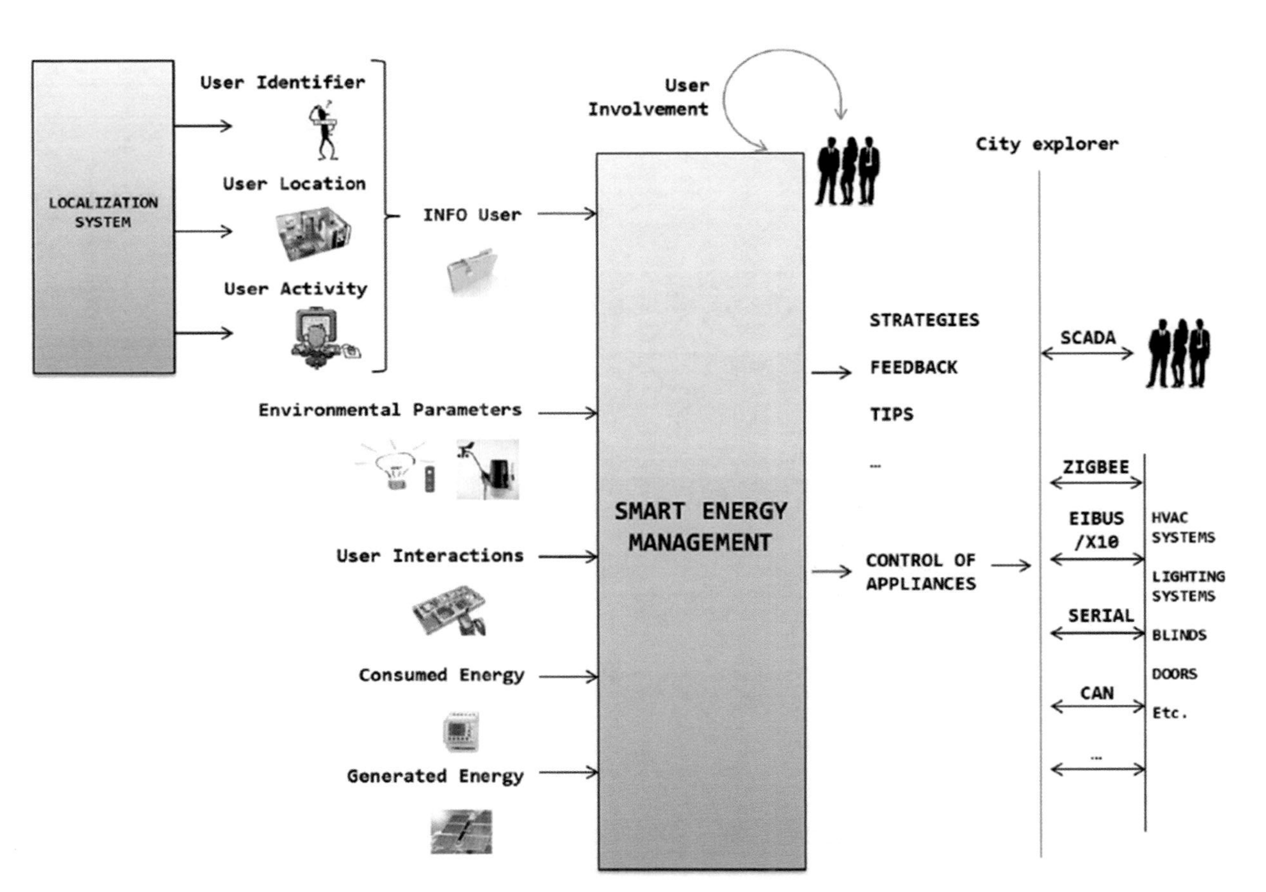

Figure 7.7 Schema of the definitive module of our building energy management system.

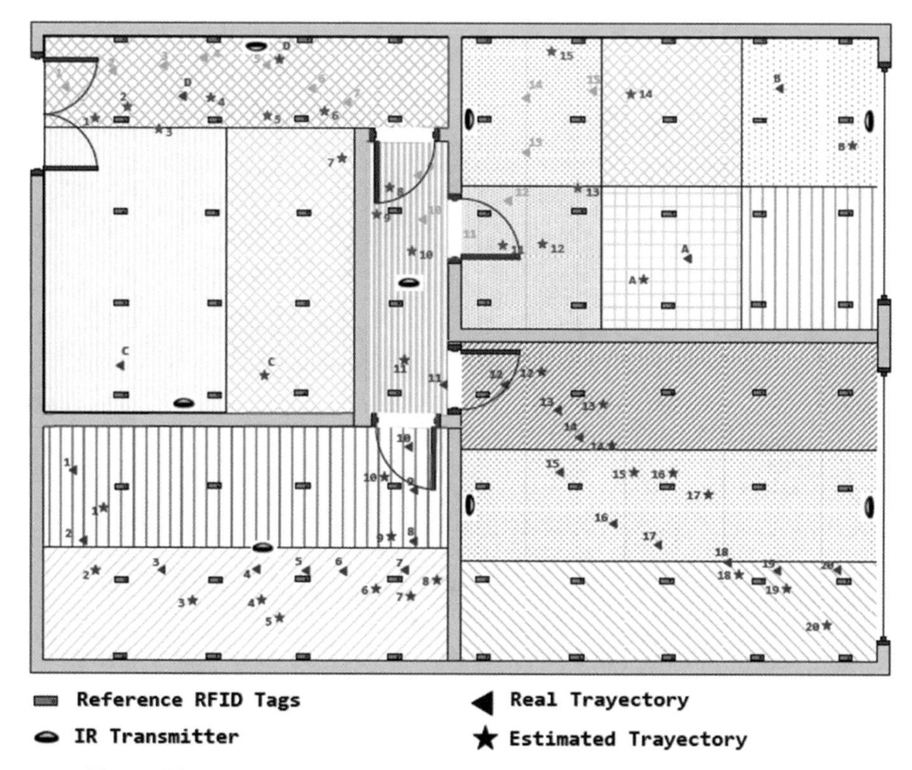

Figure 7.8 Tracking processes with a reference tag distribution of 1 m × 1 m.

7.6.2 Evaluation and Indoor Localization Mechanism

Different tracking processes are carried out in the environments considered in our tests (the TTC building) applying for this the implementation of the PF described in Algorithm 7.1. In Figure 7.8 an example of some tracking processes are carried out considering transition between different spaces of the TTC. For these paths, our system was configured to acquire data every T = 10 s. (whereas for the rest of the tests a value of T = 5 min. was considered). Taking into account the target location areas involved (represented in different colors), and the real and estimated location data provided by our mechanism, it can be safely said that it was able to monitor the user locations with a high degree of accuracy and precision.

With an 1m × 1m distribution of reference RFID tags placed on the roof of the test room, a 65% success percentage in localization is obtained having an error lower than 1 m. 98% of cases have as much 2.5 m. of error. Therefore, it can be safely said that our localization system is able to track users with a sufficient level of accuracy and precision for the location requirements associated with the comfort and energy management problem in buildings. More details about this indoor localization system can be found in [50].

7.6.3 Evaluation. Energy Consumption Prediction and Optimization

In Figure 7.9(a) it is shown the correlation heatmap between the electrical consumption of the TTC building and the outdoor environmental conditions.

It is observed that energy consumption correlates significantly ($\alpha = 0.95$) and positively with temperature, radiation, wind speed variables, vapour pressure deficit and dew point, and negatively with wind direction and humidity variables. This means that we can use safely these variables as inputs of the energy consumption model of our reference building, because they have clear impact in the energy consumption. Otherwise, precipitations are so unusual that they don't have an association with the output.

Also, a logic differentiation between situations has been considered in order to label behaviour. Situation 1: holidays and weekends, situation 2: regular mornings, and situation 3: regular afternoons. The non-parametric Kruskall Wallis test shows that energy consumption differs significantly between situations ($H(2) = 547.7$, $p < 0.01$). Also, the post hoc pairwise comparisons corrected with Holm's method retrieve a p-value smaller than 0.01, supporting the decision of creating 3 different models [51].

Thus, for each of the three situations identified for the TTC building, we have evaluated not only the punctual value of RMSE, but also we have validated whether one learning algorithm out-performs statistically significantly the others using the non-parametric Friedman test [52] with the corresponding post-hoc tests for comparison. Let xj i be the i-th performance RMSE of the jth algorithm, for this building, we have used 5-times10-fold cross validation, so $i \in \{1, 2, \ldots, 50\}$ and four techniques, so $j \in \{1, 2, 3, 4\}$. For every situation, we find significant differences ($\alpha = 0.99$) between every pair of algorithms, except for SVM and Gauss RBF ($p > 0.01$), as it is shown in Figure 7.9(b) for the particular case of situation 2.

The three models have in common that BRNN yields a better result than the other tested techniques, based on the RMSE metric. Thus, BRNN is able to generate a model with a very low mean error of 25.17 KWh – which only represents the 7.55% of the sample (this is the most accurate result) in terms of the CVRMSE. And for the worst case, BRNN provides a mean error of

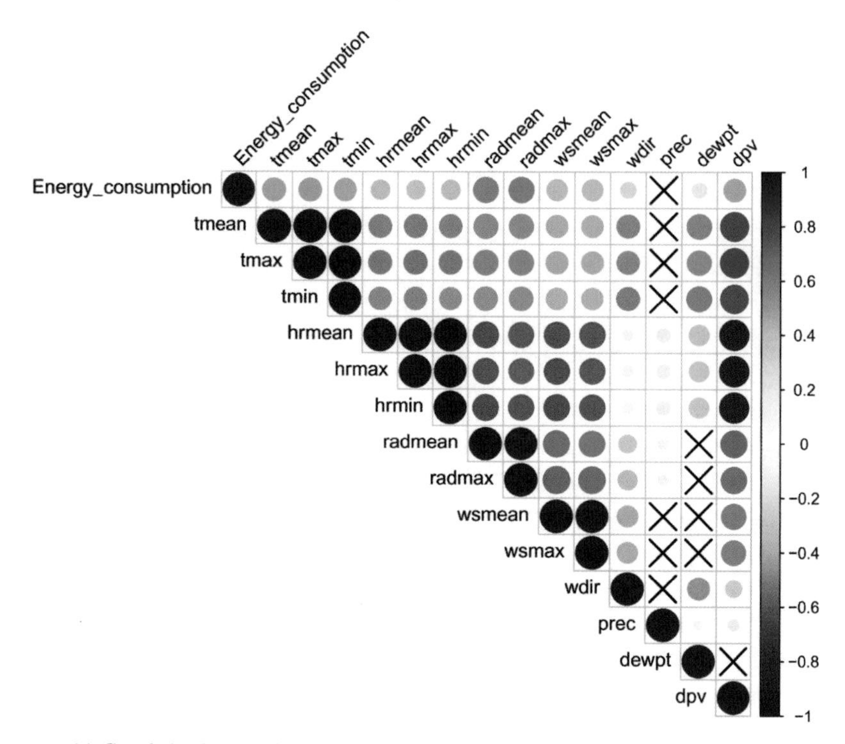

(a) Correlation heatmap betweeb consumption and outdoor environmental conditions

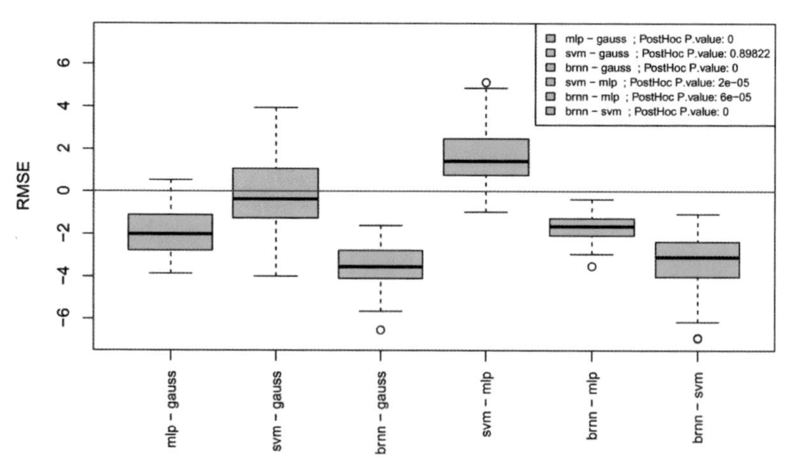

(b) Boxplots comparing models pairwise (situation 2)

Figure 7.9 Modeling results.

43.76 KWh – which represents the 10.29% of the sample in the reference TTC building – that is acceptable enough considering that the final aim is to save energy.

To evaluate our GA-based optimization strategy, controlled experiments were carried out in the TTC building with different occupant's behaviours. The results showed that we can accomplish energy savings between 15% and 31%. Trying to validate the application of our proposal we have applied it in a different scenario with limited monitoring and automation technologies, achieving energy saving of about 23%.

7.6.4 Evaluation. User Involvement

For the experiments described here, fifteen people took part in the focus group studies which help us extract user-preferences and pinpoint design concerns. Understanding user contexts, such as motivation for saving energy and the constraints for implementing energy saving behavior, enables better understanding of user preferences and how the energy monitoring system can work with users to achieve the best possible behavioral changes.

During the data collection process performed in the experiment, the subjects were asked to walk freely along the different scenarios considered, and to work or relax in the different areas designed specifically for such goals. This experiment was repeated during 3 hours per day considering different conditions of user movements and activities, environmental conditions, preferences, etc. At the time of writing, the system has just completed the first 62 days of measurement, so this time is the baseline period used to assess the impact of including users in the loop of our system. During the first 31 days of the experiment, users lacked any feedback about their energy consumption as well as any control capability over the setting of comfort and energy levels, but during the last 31 days of the experiment, users were empowered and were included as a holistic component of the system. During this second phase of the system operation, the system displayed real time energy usage in kW, cost of energy usage, energy saving tips, energy usage history (hourly, daily, monthly), etc. through both SCADA-web and the control panel installed in the target scenario. Also, during this last phase, users could define their own strategies to control any appliance or monitor any specific parameters sensed by the system.

Despite the relatively short time of evaluation (one month), a nearly analysis shows that the system has already had a positive impact on user behaviors, which can be translated into energy saving terms. Figure 7.10

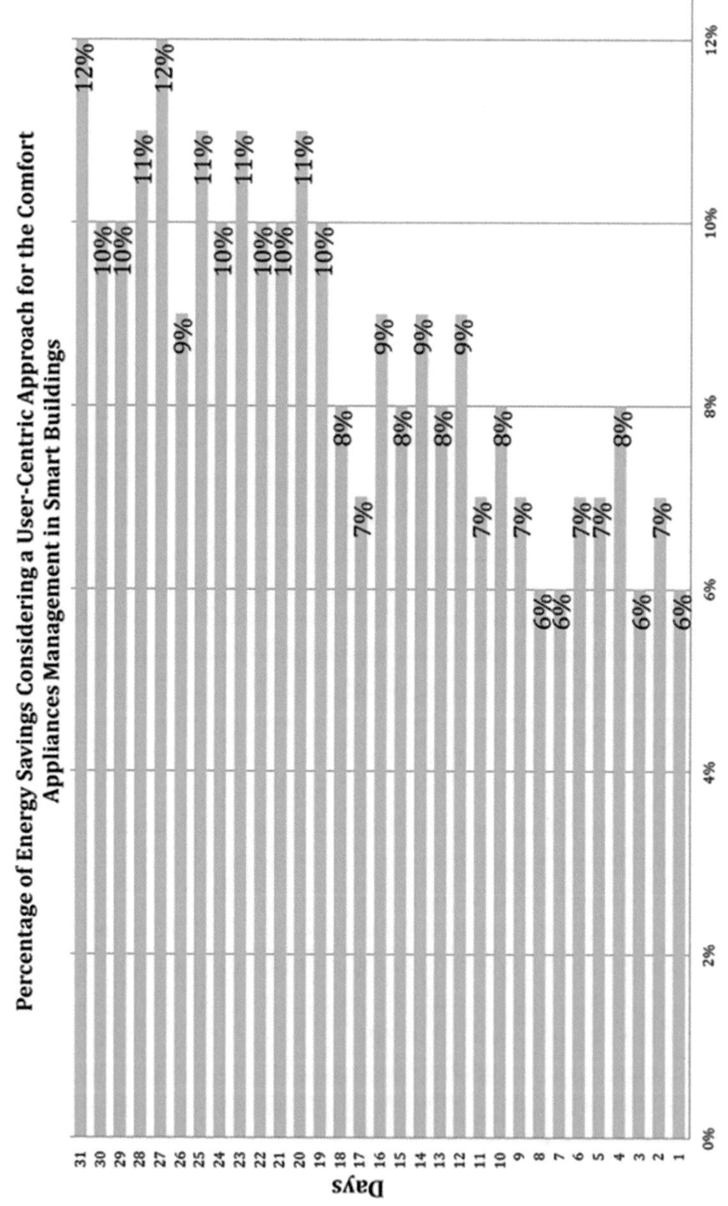

Figure 7.10 Percentage of energy consumption savings in comfort services considering a user-centric building management efficient.

shows the energy savings achieved during the second month of operation of our energy management system in contrast to the first experimental month. It can be seen how we achieved a saving of up to 12% of the energy involved, and the medium value of 9% for the experimental month. Furthermore, the results reflect how the increased savings become more stable with time, specifically from the 17th day of the system operation. The reason of this saving increasing is because our system is able to learn and adjust itself to any feedback indicated by users regarding their comfort associated profile, and to recognize patterns of user behavior.

7.7 Conclusions and Future Work

The proliferation of ICT solutions (IoT among them) represents new opportunities for the development of new intelligent services, contributing to more efficient and sustainable cities. In this sense, with the increasing urbanization seen in recent decades, there is an urgent need to achieve energy-efficient environments to ensure the energy sustainability of cities. But to achieve this goal, it is first necessary to solve energy efficiency concerns at building level, since this constitutes the cornerstone of the overall problem. For greater energy efficiency in buildings, smart solutions are required to monitor and control the capabilities offered by wide sensor and actuator networks deployed as part of the system. Furthermore, occupants play an important role in this type of system, since they are the recipients of the indoor services provided by electrical appliances installed in buildings, most of them responsible for providing them with comfort conditions. In this sense, it is required to propose building management systems able to tackle energy efficiency requirements while user comfort conditions are also taken into account. To date, however, the solutions proposed are mainly based on determinist models with few accurate predictions, and are not able to consider real-time data in most cases. Indeed, they do not even come close to reflecting reality.

In this chapter, we propose a building energy management system powered by IoT capabilities and part of a novel context and location-aware system that covers the issues of data collection, intelligent processing to save energy according to user comfort preferences and features that modify the operation of relevant indoor devices. An essential part of our energy efficiency system are the key aspects of integrating user location and identity, so that customized services can be provided to them while any useless energy consumption in the building is avoided. Furthermore, another relevant feature is users involvement with the system, through their interactions and their participation to get energy savings in the building.

The applicability of our system has been demonstrated through its installation in a reference building. Thus, using user location data, considering target regions of occupancy for comfort and energy management in the building, and finally including users in the loop of the system operation, we show that energy consumption in buildings can be reduced by a mean of about 23%. If we translate this mean value of energy saving to city level, assuming that buildings represent 40% of the total energy consumption at European level, a reduction of 9% at city level could be achieved by installing this energy management system in buildings.

The ongoing work is focused on the inclusion of people behaviour during the operational loop of this kind of systems for smart cities. Thus, for the case of smart building applications, users will be encouraged to participate in an active way through their engagement to save energy. On the other hand, in the case of the public tram service, data coming from crowd-sensing initiatives will be integrated to improve the estimation of the urban mobility patterns.

Acknowledgments

This work has been partially funded by MINECO TIN2014-52099-R Project (grant BES-2015-071956) and ERDF funds, by the European Commission through the H2020-ENTROPY-649849 and the FP7-SMARTIE-609062 Projects, and the Spanish Seneca Foundation by means of the PD program (grant 19782/PD/15).

References

[1] D. Petersen, J. Steele, and J. Wilkerson, "Wattbot: a residential electricity monitoring and feedback system," in Proceedings of the 27th international conference extended abstracts on Human factors in computing systems, pp. 2847–2852, ACM, 2009.

[2] Y. Agarwal, B. Balaji, R. Gupta, J. Lyles, M. Wei, and T. Weng, "Occupancy-driven energy management for smart building automation," in Proceedings of the 2nd ACM Workshop on Embedded Sensing Systems for Energy-Efficiency in Building, pp. 1–6, ACM, 2010.

[3] T. D. Pettersen, "Variation of energy consumption in dwellings due to climate, building and inhabitants," Energy and buildings, vol. 21, no. 3, pp. 209–218, 1994.

[4] R. Lindberg, A. Binamu, and M. Teikari, "Five-year data of measured weather, energy consumption, and time-dependent temperature variations within different exterior wall structures," Energy and Buildings, vol. 36, no. 6, pp. 495–501, 2004.

[5] S. Darby, "The effectiveness of feedback on energy consumption," A Review for DEFRA of the Literature on Metering, Billing and direct Displays, vol. 486, p. 2006, 2006.

[6] C. Fischer, "Feedback on household electricity consumption: a tool for saving energy?," Energy efficiency, vol. 1, no. 1, pp. 79–104, 2008.

[7] K. Voss, I. Sartori, A. Napolitano, S. Geier, H. Gonçalves, M. Hall, P. Heiselberg, J. Widén, J. A. Candanedo, E. Musall, et al., "Load matching and grid interaction of net zero energy buildings," 2010.

[8] C. Perera, A. Zaslavsky, P. Christen, and D. Georgakopoulos, "Sensing as a service model for smart cities supported by internet of things," Transactions on Emerging Telecommunications Technologies, vol. 25, no. 1, pp. 81–93, 2014.

[9] B. Guo, Z. Yu, X. Zhou, and D. Zhang, "From participatory sensing to mobile crowd sensing," in Pervasive Computing and Communications Workshops (PERCOM Workshops), 2014 IEEE International Conference on, pp. 593–598, March 2014.

[10] A. Llaria, J. Jiménez, and O. Curea, "Study on communication technologies for the optimal operation of smart grids," Transactions on Emerging Telecommunications Technologies, 2013.

[11] E. 15251:2006, "Indoor environmental input parameters for design and assessment of energy performance of buildings – addressing indoor air quality, thermal environment, lighting and acoustics," 2006.

[12] M. Hazas, A. Friday, and J. Scott, "Look back before leaping forward: Four decades of domestic energy inquiry," IEEE pervasive Computing, vol. 10, pp. 13–19, 2011.

[13] L. Berglund, "Mathematical modelsfor predicting thermal comfortresponse of building occupants," in Ashrae Journal- American Society of Heating Refrigerating and Air Conditioning Engineers, vol. 19, pp. 38–38, Amer Soc Heat Refrig Air-Conditioning Eng Inc 1791 Tullie Circle Ne, Atlanta, GA 30329, 1977.

[14] A. Zoha, A. Gluhak, M. A. Imran, and S. Rajasegarar, "Non-intrusive load monitoring approaches for disaggregated energy sensing: A survey," Sensors, vol. 12, no. 12, pp. 16838–16866, 2012.

[15] A. I. Dounis and C. Caraiscos, "Advanced control systems engineering for energy and comfort management in a building environment—a

review," Renewable and Sustainable Energy Reviews, vol. 13, no. 6, pp. 1246–1261, 2009.

[16] C. Ninagawa, H. Yoshida, S. Kondo, and H. Otake, "Data transmission of ieee1888 communication for wide-area real-time smart grid applications," in Renewable and Sustainable Energy Conference (IRSEC), 2013 International, pp. 509–514, IEEE, 2013.

[17] M. S. Al-Homoud, "Computer aided building energy analysis techniques, "Building and Environment, vol. 36, no. 4, pp. 421–433, 2001.

[18] D. B. Crawley, J. W. Hand, M. Kummert, and B. T. Griffith, "Contrasting the capabilities of building energy performance simulation programs," Building and environment, vol. 43, no. 4, pp. 661–673, 2008.

[19] J. Clarke, J. Cockroft, S. Conner, J. Hand, N. Kelly, R. Moore, T. O'Brien, and P. Strachan, "Simulation-assisted control in building energy management systems," Energy and buildings, vol. 34, no. 9, pp. 933–940, 2002.

[20] D. B. Crawley, L. K. Lawrie, F. C. Winkelmann, W. F. Buhl, Y. J. Huang, C. O. Pedersen, R. K. Strand, R. J. Liesen, D. E. Fisher, M. J. Witte, et al., "Energy plus: creating a new generation building energy simulation program," Energy and Buildings, vol. 33, no. 4, pp. 319–331, 2001.

[21] Z. Chen, D. Clements-Croome, J. Hong, H. Li, and Q. Xu, "A multicriteria lifespan energy efficiency approach to intelligent building assessment," Energy and Buildings, vol. 38, no. 5, pp. 393–409, 2006.

[22] V. Garg and N. Bansal, "Smart occupancy sensors to reduce energy consumption," Energy and Buildings, vol. 32, no. 1, pp. 81–87, 2000.

[23] H. Hagras, V. Callaghan, M. Colley, and G. Clarke,"A hierarchical fuzzy–genetic multiagent architecture for intelligent buildings online learning, adaptation and control," Information Sciences, vol. 150, no. 1, pp. 33–57, 2003.

[24] D.-M. Han and J.-H. Lim, "Design and implementation of smart home energy management systems based on zigbee," Consumer Electronics, IEEE Transactions on, vol. 56, no. 3, pp. 1417–1425, 2010.

[25] P. Oksa, M. Soini, L. Sydänheimo, and M. Kivikoski, "Kilavi platform for wireless building automation," Energy and Buildings, vol. 40, no. 9, pp. 1721–1730, 2008.

[26] D. O'Sullivan, M. Keane, D. Kelliher, and R. Hitchcock, "Improving building operation by tracking performance metrics throughout the building lifecycle (blc)," Energy and buildings, vol. 36, no. 11, pp. 1075–1090, 2004.

[27] G. Escrivá-Escrivá, C. Álvarez-Bel, and E. Peñalvo-ópez, "New indices to assess building energy efficiency at the use stage," Energy and Buildings, vol. 43, no. 2, pp. 476–484, 2011.

[28] V. Sundramoorthy, G. Cooper, N. Linge, and Q. Liu, "Domesticating energy-monitoring systems: Challenges and design concerns," IEEE Pervasive Computing, vol. 10, no. 1, pp. 20–27, 2011.

[29] G. Bello-Orgaz, J. J. Jung, and D. Camacho, "Social big data: Recent achievements and new challenges," Information Fusion, vol. 28, pp. 45–59, 2016.

[30] B. Pan, Y. Zheng, D. Wilkie, and C. Shahabi, "Crowd sensing of traffic anomalies based on human mobility and social media," in Proceedings of the 21st ACM SIGSPATIAL International Conference on Advances in Geographic Information Systems, SIGSPATIAL'13, (New York, NY, USA), pp. 344–353, ACM, 2013.

[31] Massive Online GeoSocial Networking Platforms and Urban Human Mobility Patterns: A Process Map for Data Collection, ch. 197, pp. 1586–1593.

[32] D. E. Difallah, M. Catasta, G. Demartini, P. G. Ipeirotis, and P. Cudré-Mauroux, "The dynamics of micro-task crowdsourcing: The case of amazon mturk," in Proceedings of the 24th International Conferenceon World Wide Web, pp. 238–247, International World Wide Web Conferences Steering Committee, 2015.

[33] C. Cardonha, D. Gallo, P. Avegliano, R. Herrmann, F. Koch, and S. Borger, "A crowdsourcing platform for the construction of accessibility maps," in Proceedings of the 10th International Cross-Disciplinary Conference on Web Accessibility, W4A'13, (New York, NY, USA), pp. 26:1–26:4, ACM, 2013.

[34] A. Piscitello, F. Paduano, A. A. Nacci, D. Noferi, M. D. Santambrogio, and D. Sciuto, "Danger-system: Exploring new ways to manage occupants safety in smart building," in Internet of Things (WF-IoT), 2015 IEEE 2nd World Forum on, pp. 675–680, Dec 2015.

[35] A. Handbook, "Fundamentals," American Society of Heating, Refrigerating and Air Conditioning Engineers, Atlanta, vol. 111, 2001.

[36] M. A. Zamora-Izquierdo, J. Santa, and A. F. Gómez-Skarmeta, "An integral and networked home automation solution for indoor ambient intelligence," Pervasive Computing, IEEE, vol. 9, no. 4, pp. 66–77, 2010.

[37] J. Santa, M. A. Zamora-Izquierdo, M. V. Moreno, A. J. Jara, and A. F. Skarmeta, "Energy-efficient indoor spaces through building automation,"

in Inter-cooperative Collective Intelligence: Techniques and Applications, pp. 375–401, Springer, 2014.

[38] H. Liu, H. Darabi, P. Banerjee, and J. Liu, "Survey of wireless indoor positioning techniques and systems, "Systems, Man, and Cybernetics, Part C: Applications and Reviews, IEEE Transactions on, vol. 37, no. 6, pp. 1067–1080, 2007.

[39] H. Ji, L. Xie, C. Wang, Y. Yin, and S. Lu, "Crowdsensing: A crowd-sourcing based indoor navigation using rfid-based delay tolerant network," Journal of Network and Computer Applications, vol. 52, pp. 79–89, 2015.

[40] A. Haug, "A tutorial on Bayesian estimation and tracking techniques applicable to nonlinear and non-Gaussian processes," MITRE Corporation, McLean, 2005.

[41] E. Standard et al., "Indoor environmental input parameters for design and assessment of energy performance of buildings addressing indoor air quality, thermal environment, lighting and acoustics," EN Standard, vol. 15251, 2007.

[42] H. Abdi and L. J. Williams, "Principal component analysis," Wiley Interdisciplinary Reviews: Computational Statistics, vol. 2, no. 4, pp. 433–459, 2010.

[43] L. Hawarah, S. Ploix, and M. Jacomino, "User behavior prediction in energy consumption in housing using bayesian networks," in Artificial Intelligence and Soft Computing, pp. 372–379, Springer, 2010.

[44] Y. Fu, Z. Li, H. Zhang, and P. Xu, "Using support vector machine to predict next day electricity load of public buildings with sub-metering devices," Procedia Engineering, vol. 121, pp. 1016–1022, 2015.

[45] M. Alamaniotis, D. Bargiotas, and L. H. Tsoukalas, "Towards smart energy systems: application of kernel machine regression for medium term electricity load forecasting," SpringerPlus, vol. 5, no. 1, pp. 1–15, 2016.

[46] R Core Team, R: A Language and Environment for Statistical Computing. R Foundation for Statistical Computing, Vienna, Austria, 2015.

[47] M. Kuhn, "Building predictive models in R using the caret package," Journal of Statistical Software, vol. 28, no. 5, pp. 1–26, 2008.

[48] J. A. Orosa, "A new modelling methodology to control hvac systems," Expert Systems with Applications, vol. 38, no. 4, pp. 4505–4513, 2011.

[49] E. Willighagen, "Genalg: R based genetic algorithm," R package version 0.1, vol. 1, 2005.

[50] M. V. Moreno, M. Zamora-Izquierdo, J. Santa, and A. F. Skarmeta, "An indoor localization system based on artificial neural networks and particle filters applied to intelligent buildings," Neurocomputing, vol. 122, pp. 116–125, 2013.

[51] J. M. Andy Field and Z. F. Niblett, Discovering Statistics Using R. Sage Publications Ltd, 1st ed., 2012.

[52] J. Demšar, "Statistical comparisons of classifiers over multiple datasets," The Journal of Machine Learning Research, vol. 7, pp. 1–30, 2006.

8

Internet-of-Things Analytics for Smart Cities

Martin Bauer, Bin Cheng, Flavio Cirillo, Salvatore Longo and Fang-Jing Wu

NEC Laboratories Europe, Heidelberg, Germany

8.1 Introduction

The Internet-of-Things (IoT) is becoming mature. It is moving from the research labs into production environments. In the initial phase, there are mainly small installations focused on a specific application, but, on the horizon, real and wide-scale deployments become visible, especially for realizing the concept of smart cities. From a few sensors providing us, for example, with weather information, we will get to large scale installations monitoring and influencing a wide range of aspects including traffic, energy, water, building infrastructures and public safety, all of which are highly relevant for the smart cities of the future. The true value of such IoT deployments can only be reached if the raw data gathered is processed and higher level information is derived that provides true insights into the real-world situation enabling humans or machines to take actions. The basis for deriving such information is provided by IoT analytics. Individual measurements, e.g., if individual cars or persons are passing a certain spot, may only provide limited benefits, but if the overall traffic situation can be derived or the behaviour of crowds can be determined, suitable actions can be taken. This requires a scalable IoT infrastructure, which can scale with the number of information sources, in particular sensors, the number of different applications and the number of users. To achieve scalability we need to look at all elements of the IoT infrastructure, from sensor nodes with limited resources, to local communication networks, gateways, networks and backend systems. Current IoT architectures typically integrate devices – using a range of different technologies–through gateways. Gateways connect the often resource constraint devices to the backend infrastructure. Information from the devices like sensor measurements are pushed into a logically centralized backend infrastructure. Cloud technologies are used for achieving scalability with respect to storage and processing power. Analytics in the backend can be provided with all the storage and processing power needed. Nevertheless, such infrastructures have their limitations:

- If the sheer amount of data overloads the network infrastructure connecting the gateways to the backend infrastructure, e.g. in the case of video cameras providing a stream of high-resolution images.
- If very fast response times are required for local actuations and the network introduces significant delays.
- If the raw data is not supposed to be stored, e.g. due to privacy information, and only processed results may be provided.
- If the frontend is provided by a different stakeholder who does not want to/is not allowed to provide the raw data.

In all these cases, IoT architectures that only support analytics in the backend are not suitable. The processing should take place in the frontend – at the edge of the network. This requires devices, gateways or specialized servers that are capable of doing the required analytics. In the case of a smart city, the IoT infrastructure needs to be able to support dynamically changing IoT devices as well as changing application requirements. In order to do so, analytic functions need to be dynamically deployed and adapted. In the following, we look at the state of the art, first with respect to the currently dominating cloud-based IoT architectures for analytics (Section 8.2) and show how analytics for crowd estimation can be supported in such a setting. Then we discuss in-depth key challenges for such architectures (Section 8.3). This is followed by a discussion of the state of the art for edge computing and a proposal for an edge-based smart city platform supporting analytics (Section 8.4). Crowd mobility is used as an example to showcase how use cases can benefit from edge-based IoT analytics. Finally, Section 8.5 provides a conclusion and an outlook on future work.

8.2 Cloud-based IoT Analytics

As first, in this section, we will investigate the cloud-based approach for IoT analytics which is the most commonly used by concrete smart cities and adopted by many projects, either funded by the European Commission or by nations, with the scope of creating smart city systems. We will describe a first example of cloud-based city platform for BigData analytics called City Data and Analytics Platform (CiDAP) and a real use-case of cloud-based data analytics such as a crowd estimation system.

8.2.1 State of the Art

Lots of efforts from both industry and academia have been made towards smart cities, but most of them focus on infrastructure construction, data collection, testbed deployment, or specific services/applications development. To support IoT analytics for smart cities, one of the key enablers is to build up a flexible and efficient big data analytics platform between connected data sources and applications. There are only a few studies already exploring big data platforms for smart cities, mainly in the Cloud environment. Examples include SCOPE [4] which is a Smart-city Cloud-based Open Platform and Ecosystem from Boston University, and FIWARE [2] which provides some building blocks for the development of a smart city platform based on the NGSI (Next Generation Service Interfaces) standard. Meanwhile, there are some ongoing projects trying to explore the opportunities and challenges of BigData for smart cities at the platform level, such as CityPulse [5], an ongoing European project exploring real-time stream processing and large scale data analytics for smart city applications. In addition, Singapore is building a new smart city platform called SmartNation [6] to enable greater pervasive connectivity, better situational awareness through data collection, and efficient sharing of collected sensor data. Several concrete smart-city architectures involving data analytics have been proposed. For example, in [19] describes the achievements of building an event driven architecture of a smart city for monitoring public area and infrastructure. All the data is seen as an event. An event can be a new measurement or a discovery of a complex event. The data coming from the Wireless Sensors Network or other subsystems (i.e. CCTV) may be filtered out or aggregated and passed to a cloud-based control center where the raw event (or almost-raw in case of aggregated data) are merged and correlated. The outcome of this processing is the creation of more and more abstracted data from less abstracted data. In case of event above certain threshold the control center would send commands back to the WSN. The analytics involved in this approach is a progressive refining of the available events till a decision. Therefore, the analysis is limited to real-time data and to very specific purposes (like event merging, event correlation or threshold checks). A similar example is described in [15], where a central reasoner is in charge of evaluating incoming aggregated data from Sensor Actuator Network

(SAN). Also, in this solution the analysis is conducted only over real-time data aggregated on the edge with the target of discovery potential critical situations. Aside from the mentioned projects above, there are industrial companies which are advertising their smart city data platforms, like IBM [25], AGT [29], Microsoft [3]. Some companies are also offering ready-to-use generic purpose IoT Platforms with embedded IoT analytics features. For example, the Amazon AWS IoT [14] is a platform capable to automatically scale in the cloud according to the load. The IoT data can be forwarded to other Amazon cloud-based services (e.g. for stream processing, for machine learning applications or for storage purposes). Another solution, more in the industrial plant context, is offered by General Electric [20]. Predix cloud offers capability to connect the gathered data from multiple Predix machines to data analytics service (time series and data analytics orchestration) and several storages options. Also, IBM offers a cloud based platform for IoT: [21]. The idea is to connect the devices or the gateways via MQTT directly to the platform. Once the data is managed in the cloud, the platform is offering integration to many services like analytics (e.g. data streaming processing, predictive analytics, geospatial analytics) and storage (e.g. SQL, NoSQL, time-series etc.).

8.3 Cloud-based City Platform

Typically, for a cloud-based smart city platform the following design issues must be taken into account: First, how to design an efficient storage system to manage a large amount of heterogeneous IoT data? Second, how to deal with both historical data and real-time data in the same platform infrastructure? Third, how to design flexible and generic application interfaces for both internal platform applications and external smart city applications? In this section we explain how these issues can be addressed in a live smart city BigData platform called CiDAP. Currently, CiDAP has been in production for several smart cities globally, such as Santander in Spain, Wellington in New Zealand, and Tokyo in Japan. The CiDAP platform is architecturally scalable, flexible, and extendable in order to be integrated with different scales of smart city infrastructures. The CiDAP platform has been deployed and integrated with a running IoT experimental testbed SmartSantander, one of the largest smart city testbeds in the world. Within the SmartSantander testbed, more than 15,000 sensors (attached with around 1,200 sensor nodes) have been installed around an area of approximately 35 square kilometer in the city. A large proportion of the sensor nodes are hidden inside white boxes and attached to street infrastructure such as street lamps, buildings and utility

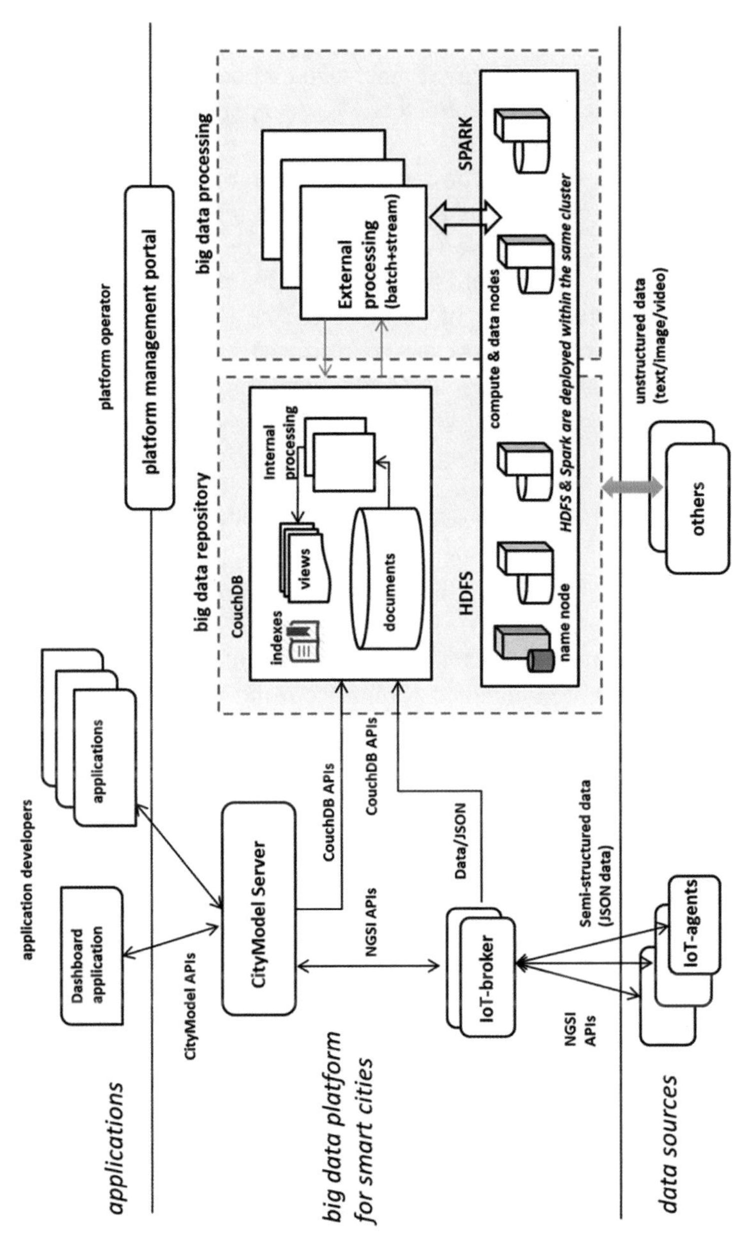

Figure 8.1 System architecture of the CiDAP platform.

poles, while others are buried into the pavement, e.g., parking sensors. Not all of the sensors are static; some are placed on the city's public transport network, including buses, taxis and police cars. The deployed sensors provide real-time information regarding different environmental parameters (light, temperature, noise, CO2), as well as other parameters like occupancy of parking slots in some downtown areas.

Here is how the CiDAP platform works at the high level (see Figure 8.1). First, data with different formats are collected via the IoT-broker [1] from multiple data sources, then forwarded to the BigData repository CouchDB, which is a document based NoSQL database. The collected data are then processed and aggregated by a set of pre-defined or newly launched processing tasks. The simple processing tasks can be performed by the BigData repository internally, such as transforming data into new formats or creating new structured views/tables to index data. Any complex or intensive processing tasks, such as aggregating or mining data via advanced data analytics, must be separated from the BigData repository so they can be efficiently and externally executed over the BigData processing module, which provides more flexible and scalable computation resource based on a Spark [12] cluster with a large number of compute nodes. Since the BigData repository can already handle lightweight processing tasks in a scalable and incremental manner, the BigData processing module can be optional if we do not need intensive data processing or analytics. By fetching generated results from the BigData repository or forwarding messages directly from data sources in the smart city testbed, a CityModel server is designed to serve queries and subscriptions from external applications based on pre-defined CityModel APIs. Meanwhile, a web-based platform management portal is provided to the platform operator to monitor the status of the entire BigData platform.

All external applications communicate with the CiDAP platform via the CityModel server based on a REST based API, called CityModel API. The CityModel API allows application to do simple query, complex query, and subscription. A simple query requests aggregated results over the latest status of all sensors, which represent the latest and real-time snapshot of the entire city testbed, while a complex query can request aggregated results over the historical data collected within a specified time range. Subscription is the mechanism to keep applications always notified with the latest results so that the application does not have to query the data all the time. There are two types of subscriptions, CacheDataSub and DeviceDataSub, as illustrated by Figure 8.2. The difference is CacheDataSub goes to the data repository CouchDB while DeviceDataSub goes directly to physical

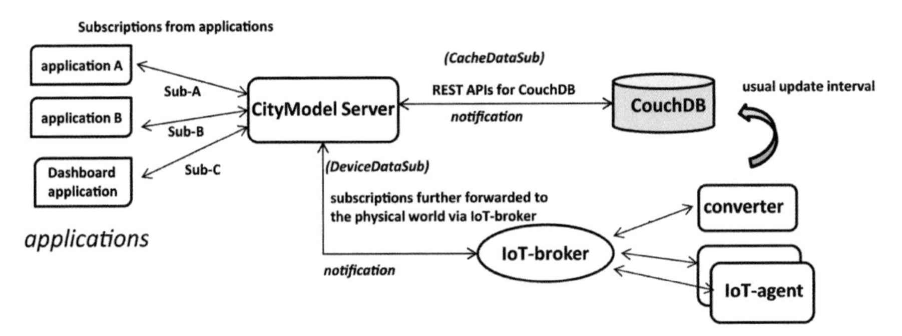

Figure 8.2 Subscription mechanisms to get real-time notifications.

devices in the IoT testbed through IoT-broker and IoT-agents. Both of them are designed to notify applications with real-time changes, but with different expected latency. The notification latency for CacheDataSub is relatively longer than the one for DeviceDataSub, because devices will fire notifications immediately after the requested changes happen, without waiting for the next report period. Unfortunately, the DeviceDataSub is not fully working in the integrated system with the SmartSantander testbed because the sensor nodes in this testbed can only report updates in a passive and periodic way.

CiDAP is just a concrete example to illustrate how a smart city platform could be designed to support flexible IoT analytics in a cloud environment. On the other hand, based on our experiments and experience with CiDAP, we have also identified certain limitations of the cloud-based solution. For example, with the cloud-based solution it is difficult to support time-critical use cases such as autonomous driving and real-time emergency detection, because the responsive time to react on real-time situation could be more than 10 seconds. However, this limitation can be overcome by edge-based solutions, which will be introduced in Section 8.3.

8.3.1 Use Case of Cloud-based Data Analytics

The crowd estimation system is to map a set of sensing readings into a certain level of crowd density. Figure 8.3 shows the system overview of cloud-based crowd estimation proposed in [23]. The system is deployed in a shopping mall, where 23 sensors are installed. The size of the shopping mall is roughly 90 square meters. The sensors continuously report ambient information (such as CO2, noise level, temperature, and humidity) to the

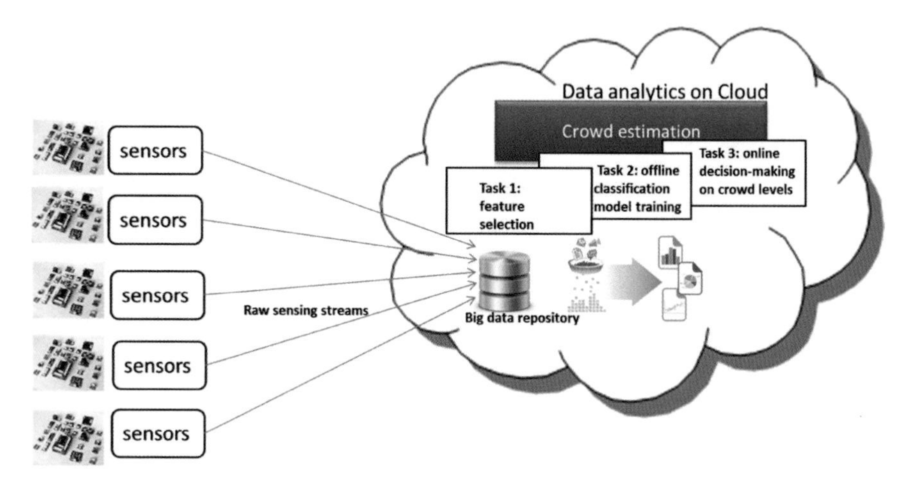

Figure 8.3 System overview of the cloud-based analytics.

BigData repository. In this system, all tasks of data analytics are handled by Cloud. The task of feature selection chooses the most important information behind the sensing data which will be the input of the task 2. For each sensor, the task extracts the mean, standard deviation, variance, minimum and maximum from sensing readings. In addition, the slope of temperature, humidity, noise and CO_2 readings are considered. The second task trains a classification model based on the features selected by task 1, where this task considers four different supervised learning algorithms including Naïve Bayes Classifier [31], C4.5 [26], Random Forests [17], and Support Vector Machines [18]. The ground truth is established through human observations, where the system pre-defines 4 crowd levels from 0 to 3 which are mapped to "occasional passer-by", "sparse traffic", "crowded" and "overcrowded" respectively.

Based on the observations by the building management office, the crowd level 0 is mapped to 0–15 people, the crowd level 1 is mapped to 16–30 people, the crowd level is mapped to 31–45 people, and the crowd level 4 is mapped to more than 45 people. Finally, the third task performs decision-making to estimate levels of crowd density when real-time sensing data arrives. Given a location and features from multiple types of sensing readings, this task can map those information to a level of crowd density. However, IoT data contains much useless and redundant information such as zero readings. Meanwhile, an IoT platform may serve many IoT applications simultaneously and some of real-time IoT applications may have critical QoS requirements. To support real-time applications, flexible and dynamic data analytical models across the system will be preferred, where some processing tasks can be offloaded onto edge.

8.4 New Challenges towards Edge-based Solutions

Different from traditional data analytics like Web analytics and log analytics, IoT analytics must deal with the following IoT system characteristics:

1. IoT data are usually unstructured stream data and constantly generated from geo-distributed sensors over time, ranging from time series event streams to high data rate video streams; sending all raw data to the centralized Cloud for further processing will be very costly and also introduce too much traffic to the underlying network;
2. Mobility and co-location of sensors and actuators, meaning that both sensors and actuators are possible to move and actuators usually require data from nearby sensors;
3. Actuators often expect low latency results to make fast actions;
4. Raw data and derived results are also expected to be shared and consumed across different parties from anywhere, either globally from the Cloud or locally from a nearby region, because the cost to deploy the infrastructure of a large-scale IoT system could be very high and the platform and the sensor data are worth to be shared for maximizing their benefits. All of these requirements bring new technical challenges to IoT analytics since problems like data distribution, data reliability, real-time data processing, processing flexibility, and platform openness need to be considered and addressed differently.

Regarding the requirements of IoT analytics, there is currently a new trend to move processing to the edge, where IoT data are generated and analytics results are consumed. Traditional computing models collect IoT data and then transmit them to a data center for doing scalable data analytics, but this is no longer a sustainable and suitable model for large-scale IoT systems.

Our previous experimental results from CiDAP also indicate that some processing should be shifted from the Cloud down to the edge or IoT devices, especially when applications expect to have real-time analytics results within a few seconds or even a sub-second. Since IoT data are not only big, but also naturally geo-distributed and increasing over time, processing all data only in the Cloud will introduce high bandwidth cost between the network edges and the Cloud. In many cases, it would make more sense to process or compress data before transmitting the data to the Cloud, or transmit only selected data or derived results (e.g., anomalies, exceptions, averages). Therefore, for large IoT systems there is a strong need to do analytics at the network edges. For edge-based IoT analytics, the following challenges must be taken into account.

1. *Scalability*: Edge-based IoT analytics needs to be able to scale up to thousands of geo-distributed nodes over the wide area network. For example, if we consider IoT gateways and even users' mobile phones as edge nodes, the total number of edge nodes in a large scale IoT system can be easily over 1000. According to the recent report by Yahoo [7], supporting over 1000 nodes with Storm within a cluster is already problematic due to the bottleneck of its zookeeper service component.
2. *Task Optimization*: A sophisticated task scheduling algorithm is required to optimize the resource usage and minimize the latency to deliver analytics results. The underlying network topology of all IoT agents needs to be considered by the task scheduling algorithm as well, since it can affect the latency and the bandwidth consumption to produce analytics results. Also, the task scheduling algorithm needs to be aware of the geo-locations of sensors and actuators.
3. *Flexible Application Interfaces*: application developers should have enough freedom to implement their own processing tasks for any type of streams, such as event streams, text streams, and video streams. Further, they should be able to define their application requirements and to access real-time analytics results from their applications. Although this can be built on top of existing solutions, none of the latter includes inherent platform interfaces for supporting this.
4. *Multi-tenancy Support*: The designed edge analytics platform should allow multiple users to share the same edge analytics infrastructure by ensuring efficiency, fairness, and quality of service. This must be achieved by designing sophisticated task scheduling and resource orchestration mechanisms. Resource sharing across applications and users is highly important, since the deployment and maintenance cost for a large-scale IoT system is still big and its value should be maximized by enabling more sharing across various applications and users.
5. *Openness and Security*: Edge-based IoT analytics platform is provided as a PaaS for a set of IoT applications to do stream-based edge analytics. Therefore, it will be important to consider the openness and security issues at the design phase.

8.5 Edge-based IoT Analytics

In this section we examine the edge-based approach for data analytics, which is still at a very early stage in the Smart City context and in general in the Internet-of-Things world. We will introduce our edge-based solution for IoT analytics, describing the architecture in every components and the system workflow. Also for this solution we provide a real-use case of IoT analytics applied to our edge-based solution.

8.5.1 State of the Art

Fog computing is a term first advertised by Cisco, also known as Edge Computing [30]. Basically it refers to extending cloud computing to the edge by allowing data processing to happen at the network edges. As reported by the survey of [16], fog computing has been introduced mainly because of the strong needs of IoT systems for low latency results and fast decision makings. Cisco has created a platform called IOx to support fog computing by hosting applications in a Guest OS running in a hypervisor directly on the network routers. Like a virtual machine, IOx enables running scripts or even compiled code at the network edge. Although fog computing providers like Cisco establish an environment to do distributed computation at edges, to benefit from such environment enterprises still need a system that determines which data needs to be processed immediately at the edge and which data should be moved to the Cloud for further deep analysis. Currently, as compared with cloud computing, fog computing is still in the very early stage and lacks sophisticated data analytics platforms that allow us to efficiently utilize the power of the edges and the Cloud together.

As a new trend, edge-based IoT analytics aims to leverage the power of both fog computing and cloud computing to support real time stream processing. Only a few early stage studies have been done in this area, for example, a recent work from Carnegie Mellon University [27] proposes a VM-based edge computing platform for performing video analytics at the network edges, but it only focuses on video streams and does not consider how to define topologies to do customized stream processing on top of the edges and the Cloud. In addition, some industrial systems have been done to explore edge analytics, such as AGT IoT analytics platform [8], Geo-distributed analytics from ParStream [11], and Quarks from IBM [13]. However, the details of their system designs are not open. From what they advertise about their systems, none of them seems to support multi-tenancy and dynamic topology execution.

The usage of edge computing in concrete smart cities deployment is usually meant only for data aggregation [19] or for semantic reasoning on local data [15]. The computation procedures are statically installed on the edge node and only pre-defined commands (like threshold settings) can be sent by the central application.

8.5.2 Edge-based City Platform

To address the above challenges, we introduce our new edge-based city platform called Geelytics in this section. As an edge-based platform solution for IoT analytics, Geelytic is not supposed to be a replacement of the cloud based solution like CiDAP, but rather an enhancement or a complementary solution to relax the pure cloud-based solution with the capability of edge analytics.

Geelytics is mainly designed for large scale IoT systems that consist of a large number of geo-distributed data producers, result consumers, and compute nodes that are located both at the network edge and in the cloud. Data producers are typically sensors, connected cars, glasses, video cameras, and mobile phones, being connected to the system via different types of edge networks (e.g., Wi-Fi, ZigBee, or 4G, but maybe also fixed networks). They are constantly reporting heterogeneous, multi-dimensional, and unstructured data over time. On the other hand, result consumers are actuators or external applications that expect to receive real-time analytics results from sensor data and then take fast actions accordingly. Both data producers and result consumers could be either stationary or mobile. In between them there are lots of compute nodes geographically distributed at different locations. In general compute nodes are heterogeneous in terms of resource and data processing capability and they can be located at different layers of the network. For example, they could be small data centers at base stations in a cellular network or IoT gateways in factories or shops.

The Geelytics system is designed as an IoT edge analytics platform that allows consumers to dynamically trigger certain stream data processing either at network edges or in the Cloud to derive real-time IoT analytics results from a set of data providers. At very high level, it works like a distributed pub/subsystem to interact with geo-located sensors and actuators, meanwhile having a built-in stream processing engine that can perform on-demand IoT stream data analytics based on the underlying Cloud-Edge system infrastructure. As shown in Figure 8.4, the Geelytics platform includes the following components.

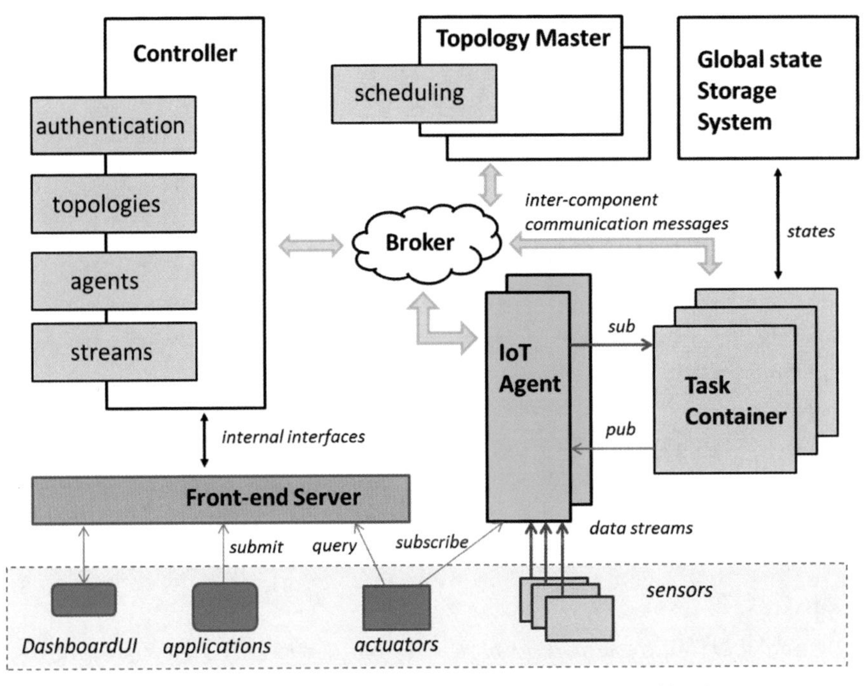

Figure 8.4 Subscription mechanisms to get real-time notifications.

IoT Agent: each IoT agent represents one worker that is capable of performing stream processing tasks. In Geelytics IoT agents are deployed on geo-distributed physical or virtual machines, either in a cluster in the Cloud or at the network edges. Each agent communicates with the Controller to report its capability and available resources, accepts incoming tasks from the topology masters, and instantiates them locally. It can also receive data streams from nearby sensors and fetch data streams from other remote IoT agents according to the requirements from its accepted running tasks. All IoT agents have the same role, but they might be heterogeneous depending on the processing capabilities and network connections of their host.

Task Container: every schedulable task is wrapped up as an application container by developers. Based on Docker [9], it can be fast deployed and executed anywhere by an IoT agent. By design, each running task within an application container will communicate with its IoT agent via a pub/sub mechanism, including subscribing input streams and publishing generated output streams. Using Docker as the environment to run IoT analytics tasks, we are able to better support multi-tenancy, because Docker allows us to do fine-grained resource allocation for each task.

Topology Master: In Geelytics, an IoT analytics application consists of a set of correlated data stream processing tasks. Each application has a dedicated topology master to manage all involved stream processing task instances running in the Cloud or at the network edges. Each topology master is responsible for monitoring and allocating tasks to different IoT agents. It requests the current state of all available resources, including all active IoT agents and their remaining capabilities, network latency and traffic across IoT agents, and then make decisions on at which IoT agent each task must be instantiated, regarding the task topology specification and optimization objectives given by the application developer and the current workload. By separating Topology Master from the Controller, Geelytics is able to achieve better scalability as compared with existing stream processing platforms like Storm. In addition, several task assignment algorithms have been applied by Topology Master to optimize task allocation between Cloud and edges during the runtime, with regards to the objectives of reducing bandwidth consumption and latency.

Controller: all system resources and core components are managed by the Controller, which is a single central control point of Geelytics running in the Cloud. It indexes all streams, agents, topologies, and users. For security reasons, it authenticates all the other components, especially IoT agents, when they join the system.

Front-end Server: application interfaces are supported by the front-end server via HTTP REST, enabling that: 1) application developers can submit the task definition, topology structure, and optimization objectives; 2) actuators can query or subscribe the analytics results generated by the submitted application; 3) sensors can register them to a nearby IoT agent; 4) a dashboard service is provided to check the status of the entire Geelytics system and also to manage users and applications.

Broker: the Broker is a distributed message exchange system to enable the communication between different components. To be scalable and flexible, the Broker must have high throughput and support topic-based message handling.

Global State Storage System: Geelytics is designed to support complex stream processing tasks, such as machine learning tasks, image or video processing tasks. For those tasks, it is important to save some of the intermediate states to tolerate unexpected failures. The same concern goes for the other components as well, such as the Controller, IoT agents, and the topology master. Therefore a global state storage system is introduced to keep all intermediate states and results, using existing NoSQL database systems such as key-value based Redis or document-based CouchBase.

8.5.3 Workflow

The system platform is initialized in the following sequence. First, the broker and the global state storage system must be set up independently as two external sub-systems. Then the Controller is started in the Cloud and it will launch the front-end server. IoT agents can be started before or after the controller, but each of them must be authenticated by the controller when they join the system. As stream data sources, sensors can be attached to an IoT agent manually or be forwarded to the nearby IoT agent by the controller when they join the system. After the system is ready, developers need to register a user account and then start to submit their customized tasks and application topologies. Once a new application is submitted, the Controller will allocate proper resources for the application according to its requirements and then return a URL address to the actuators of this application for accessing the analytics results.

8.5.4 Task and Topology

As the example in Figure 8.5 shows, in Geelytics a data analytics application is defined by a task topology, which specifies the relationship between different stream processing tasks within the application. Based on the task topology, a processing topology will be created on the fly to handle the current workload. The processing topology consists of a set of running task instances, allocated by the topology master to the network edges or the Cloud, up to where the involved data sources are located and where the results are demanded by the actuators. In Storm a processing topology is constructed when the task topology is submitted, according to the parallelism of each task defined by the developer. In contrast, in Geelytics all data streams generated by each task in the task topology are accessible to actuators and the processing topology is constructed and changed as actuators join and leave.

In Geelytics the way to implement a task is flexible. A task just needs to follow a pub/sub communication interface to fetch the input streams and publish the output streams and a set of parameters are configured with the task to decide which input streams to bring in. However, how to handle the input streams within the task is a black box for Geelytics. Developers can use

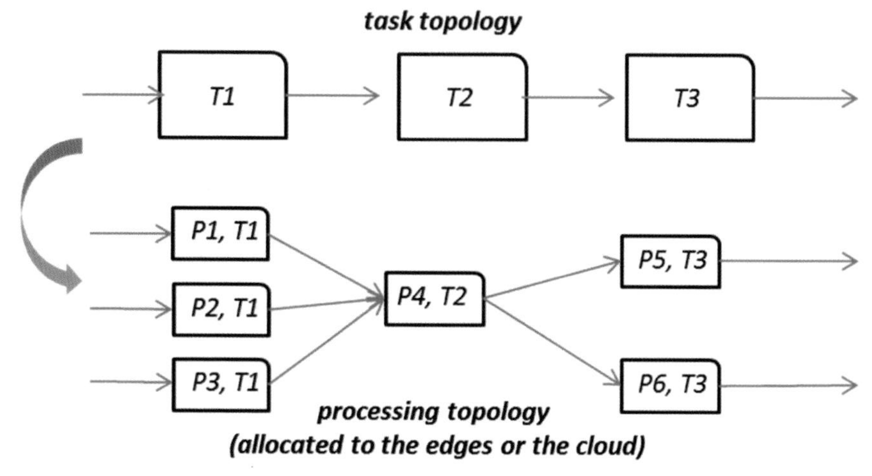

Figure 8.5 Task topology and processing topology.

any image/video processing or machine learning library to implement their tasks in any programming language, because they can wrap up all dependent libraries into a single Docker container image. In addition, all tasks can use the interface of the global state storage system to save or retrieve their state data.

8.5.5 IoT-friendly Interfaces

Geelytics also provides friendly interfaces for both data producers and result consumers to interact with the system. In Geelytics all data producers report their availability, profiles, and data streams to the system, managed by a repository based on ElasticSearch [10]. The way to fetch the data streams generated by data producers can be push-based or pull-based. In the push-based approach, data producers publish their stream data to the MQTT broker on the nearby compute node; while in the pull-based approach, data producers just announce the URLs of their streams, and later on it is up to task instances to fetch the data directly. A data producer first has to ask the controller to find a nearby worker and then registers its data stream via the nearby worker with the following details: its device ID and location, the generated stream type, and the manner to provide the stream data (push-based or pull-based). A unique ID will be returned to the producer as the global identity of its data stream. If the stream is pull-based (for example, a web camera), a URL must be provided for accessing data as well; if the stream is push-based, using the unique ID as the topic, the producer publishes the generated data to the broker provided by the nearby worker via MQTT.

A result consumer also needs to ask the controller to find a nearby worker first. After that, it sends a scoped subscription to the nearby worker for triggering some real-time data processing over the specified data sources. A subscription ID is returned to the consumer to make a further subscription to the broker. The consumer can receive the subscribed results as soon as they are produced by the triggered task instances in Geelytics. Those running tasks will be terminated once the consumer decides to unsubscribe to the result or its leaving without notice has been detected.

8.6 Use Case of Edge-based Data Analytics

A real use-case for edge-based data analytics is the study of crowd mobility pattern analysis and prediction. In the next subsections we are going to examine how we can design such application and how it is fitting in the Geelytics platform.

8.6.1 Overview of Crowd Mobility Analytics

Crowd mobility analytics investigate how many people in a certain area and how they move from one area to the others which can provide insights for various IoT applications. For example, stadium operators may need to know the number of people in a big event in case of emergencies, and airport or public transport operators may need to know passenger flows for predictable service enhancement and maintenance scheduling.

Figure 8.6 shows the overview of the crowd mobility analytics system, where the IoT platform consists of data sources, edge nodes, and cloud nodes. The data sources include Wi-Fi sensing stations and ambient sensors. Each Wi-Fi sensing station captures Wi-Fi probe requests broadcast by mobile devices from time to time, while ambient sensors include CO2, noise, temperature, humidity, and motion sensors which capture the influence of human mobility on the environment. Each edge node serves as a local data aggregator to connect to cloud nodes. Cloud nodes cooperate with edge nodes though shared backed database. Since the architecture of edge-based data analytics provides a more flexible task processing topology, it opens up more opportunities for processing data streams in a pipeline way which can speed up data analytics. Thus, the crowd mobility analytics decompose the mobility data analytics into six processing tasks based on the dependency among tasks:

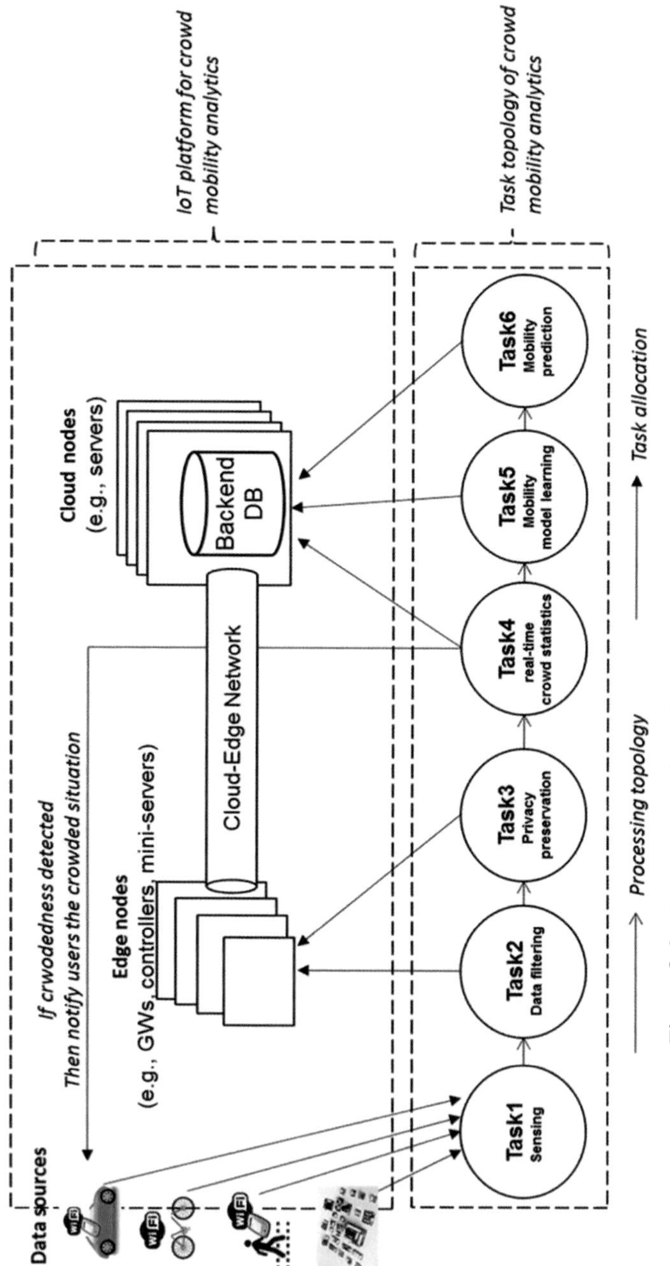

Figure 8.6 A system overview of the edge-based data analytics for crowd mobility.

(1) Wi-Fi sensing, (2) data filtering, (3) privacy preservation, (4) real-time crowd statistics, (5) mobility model learning, and (6) mobility prediction. The first three ones are lightweight sensing and data pre-processing tasks which will be assigned to edge nodes, while the latter three tasks are computation-intensive which will be assigned to cloud nodes.

8.6.2 Processing Tasks and Topology of Crowd Mobility Analytics

Since edge nodes have enough capabilities to run simple routine, the processing topology of crowd mobility analytics is designed to decouple the computation complexity between cloud and edge nodes.

First, we introduce the three lightweight tasks allocated to edge nodes as follows.

- Task 1: sensing. We build passive sensing stations to capture Wi-Fi packets broadcast by mobile devices and sensing readings, where each sensing station was build using a Raspberry Pi 2 with Arch Linux, a Wireless LAN USB Adapter, and ambient sensors.
- Task 2: data filtering. Since the previous sensing task captures all of Wi-Fi packets including dense beacons, this task picks up only Wi-Fi probe requests and represents sensors readings as a common format. Meanwhile, this task transforms the raw sensing data into a structured format for further mobility analysis in cloud nodes. For each Wi-Fi probe request packet, the system extracts the BSSID, the Wi-Fi channel on which the packet has been sent, the source and destination MAC addresses, the time when the packet has been detected, and the Wi-Fi device vendor inferred from the first 3 bytes of the MAC address.
- Task 3: privacy preservation. To avoid exposing identities of mobile users, edge nodes perform MAC address anonymization for the privacy preservation purpose. Thus, each edge node sends hashed MAC addresses to the backend database using a SHA-1 algorithm [22].

Afterwards, cloud nodes perform the following three key tasks: real-time crowd statistics, mobility model learning, and mobility prediction.

- Task 4: real-time crowd statistics. This task performs feature extraction and statistical analysis based on the results from many instances of Task 3 in a real-time way. The extracted features include the mean, maximum, minimum, and standard deviation of sensing readings from ambient sensors. The statistical analysis results include the number of mobile devices, distribution of device brands, and the number of mobile devices moving from one sensing area to the others based on the Wi-Fi.

- Task 5: mobility model learning. Based on these features extracted in Task 4, this task trains a classification model to estimate the number of mobile users in a certain area.
- Task 6: mobility prediction. Based on captured Wi-Fi probes, the task models human mobility as a finite Markov Chain [28] which represents mobility behaviour of public crowds instead of focusing on each individual's mobility trajectories. The behavioural characteristics of crowd mobility can be mapped to a level of crowd which explains how many people staying in a certain area. Furthermore, we can use this model to predict crowd levels based on the statistical analysis of mobility flows among multiple areas.

8.7 Conclusion and Future Work

In this article we discuss the technical challenges to support flexible IoT analytics for smart cities from a platform perspective. As the scale of IoT devices in a smart city is fast growing and fast response time is highly demanded by more and more smart city use cases, for IoT analytics there is a new technology trend to move data processing from the cloud to the network edges. With two concrete platform examples, namely CiDAP and Geelytics, we illustrate this new technology trend and show use cases can benefit from them.

For the time being, CiDAP focuses more on the cloud side while Geelytics focuses more on the edge side. However, Geelytics is not supposed to replace CiDAP as an alternative solution, but rather enhance it as a complementary solution. For example, Geelytics is good at processing stream data both in the cloud and at edges, but it is not a good choice for dealing with large scale historical data in the cloud, which is the strength of CiDAP. Therefore, it makes sense to integrate CiDAP and Geelytics to have a more advanced and unified platform for IoT analytics, which can utilize both edge computing and cloud computing. This is one of the future steps in the short term. In addition, we are further working on the task assignment algorithms in Geelytics to support mobility aware IoT analytics for moving objects, such as connected cars and flying drones.

In the long term, we are working on the issue of semantic interoperability to support advanced IoT analytics that can utilize the data from various data sources across different application domains. In a smart city, relevant data could come from various data sources, either in the same IoT system, from other IoT systems or even from more traditional IT systems whose content may be entered by humans. Semantic interoperability will allow us to interact with various data sources with ensured consistency of the data across systems regardless of individual data format. The semantics can be explicitly defined using a shared vocabulary as specified in an ontology.

For IoT to be successful, standardized solutions are needed – be it formal standards or de-facto standards developed as part of industry alliances or open source communities. In CiDAP we are making use of the OMA NGSI Context interfaces that are at the core of FIWARE [2] Platform. We are also actively participating in the oneM2M [24] standardization. Ultimately, important functionalities developed and explored in our research prototypes need to become part of standardization. Different standards have to be aligned and gaps in standardization have to be identified and closed.

Regarding semantic interoperability, we have integrated basic semantic functionality into oneM2M. Based on this we have done an experiment to show how semantic information can be used for converting IoT data in oneM2M into the NGSI data format used in FIWARE. We are now planning to generalize the approach using the concept of mediation gateways.

For the future work, we would also like to consider the security and privacy issues in IoT analytics for smart cities. We have done some work to ensure the secure communication between different components in both CiDAP and Geelytics, but this is still the basic step to ensure security. With the support of edge analytics, the IoT analytics platform is now geographically deployed with the extension further down to the edges, like mobile base stations, IoT gateway, and even some endpoint devices as well. In this case, it is becoming more challenging to secure the platform and IoT data.

For example, some intrusion detection might be needed to detect attacks and potential threats in real time. The research on privacy in IoT is still at an early stage, but will be an essential point for the adoption of IoT on a large scale.

References

[1] Aeron open source project. https://github.com/Aeronbroker/Aeron/
[2] FIWARE Open Source Platform. http://www.fi-ware.org/
[3] Microsoft CityNext Solution. http://www.microsoft.com/global/en-us/citynext/RichMedia/Safer_Cities/CityNext_Brochure_Safer_Cities_SML_FY15.pdf

[4] SCOPE: A Smart-city Cloud-based Open Platform and Ecosystem. http://www.bu.edu/hic/research/scope/
[5] The citypulse project, 2014.
[6] Singapore smart nation platform, 2014.
[7] The Evolution of Storm at Yahoo and Apache. http://yahoohadoop.tumblr.com/post/98751512631/the-evolution-of-storm-at-yahoo-and-apache/, 2015.
[8] AGT IoT Analytics Platform. https://www.agtinternational.com/iot-analytics/iot-analytics/analytics-the-edge/, 2016.
[9] Docker. https://www.docker.com/, 2016.
[10] ElasticSearch. https://www.elastic.co, 2016.
[11] ParStream Geo-distributed Analytics. https://www.parstream.com/product/parstream-geo-distributed-analytics/, 2016.
[12] The Apache Spark project. https://spark.apache.org, 2016.
[13] The Quarks Project. http://quarks-edge.github.io, 2016.
[14] Amazon. Amazon AWS IoT. https://aws.amazon.com/iot/.
[15] A. Attwood, M. Merabti, P. Fergus, and O. Abuelmaatti. Sccir: Smart cities critical infrastructure response framework. In Developments in E-systems Engineering (DeSE), 2011, pages 460–464, Dec 2011.
[16] Flavio Bonomi, Rodolfo Milito, Jiang Zhu, and Sateesh Addepalli. Fog Computing and Its Role in the Internet of Things. In Proceedings of the First Edition of the MCC Workshop on Mobile Cloud Computing, pages 13–16, 2012.
[17] L. Breiman. Random forests. Mach. Learn. 2001.
[18] C.-C. Chang and C.-J. Lin. Libsvm: A library for support vector machines. In ACM Trans. Intell. Syst. Technol., 2011.
[19] L. Filipponi, A. Vitaletti, G. Landi, V. Memeo, G. Laura, and P. Pucci. Smart city: An event driven architecture for monitoring public spaces with heterogeneous sensors. In Sensor Technologies and Applications (SENSORCOMM), 2010 Fourth International Conference on, pages 281–286, July 2010.
[20] General Electric. GE Predix. https://www.predix.com/.
[21] IBM. IBM IoT FouT. D. Pettersen, "Variation of energy consumption in dwellings due to climate, building and inhabitants," Energy and buildings, vol. 21, no. 3, pp. 209–218, 1994.
[22] IETF-RFC. Us secure hash algorithm 1 (sha1). https://tools.ietf.org/html/rfc3174, 2001.
[23] Salvatore Longo and Bin Cheng. Privacy preserving crowd estimation for safer cities. In Proceedings of the 2015 ACM International Joint

Conference on Pervasive and Ubiquitous Computing and Proceedings of the 2015 ACM International Symposium on Wearable Computers, pages 1543–1550. ACM, 2015.

[24] oneM2M. oneM2M – Standards for M2M and IoT. http://www.onem2m.org/aboutonem2m/why-onem2m

[25] M. Kehoe P. Fritz and J. Kwan. IBM Smarter City Solutions on Cloud. 2012.

[26] J. R. Quinlan. C4.5: Programs for Machine Learning. Morgan Kaufmann Publishers Inc., San Francisco, CA, USA, 1993.

[27] Mahadev Satyanarayanan, Pieter Simoens, Yu Xiao, Padmanabhan Pillai, Zhuo Chen, Kiryong Ha, Wenlu Hu, and Brandon Amos. Edge Analytics in the Internet of Things. IEEE Pervasive Computing, 14:24–31, June 2015.

[28] J. Laurie Snell. Introduction to Probability. Random House, New York, 1988.

[29] Martin Strohbach, Holger Ziekow, Vangelis Gazis, and Navot Akiva. Towards a big data analytics framework for iot and smart city applications. In Modeling and Processing for Next-Generation Big-Data Technologies, pages 257–282. Springer, 2015.

[30] Shanhe Yi, Cheng Li, and Qun Li. A Survey of Fog Computing: Concepts, Applications and Issues. In Proceedings of the 2015 Workshop on Mobile Big Data, pages 37–42, 2015.

[31] H. Zhang. The optimality of Naive Bayes. In Proceedings of the Seventeenth International Florida Artificial Intelligence Research Society Conference, 2004.

9

IoT Analytics: From Data Collection to Deployment and Operationalization

John Soldatos and Ioannis T. Christou

Athens Information Technology, Greece

9.1 Operationalizing Data Analytics Using the VITAL Platform

The VITAL smart cities platform has been introduced in an earlier chapter (Chapter 4). It comprises a set of middleware libraries and accompanying tools, which facilitate the development, deployment and operation of smart cities applications, including IoT analytics applications. The platform supports functionalities across all the phases of the IoT analytics lifecycle, which have been presented in the introductory chapter. The rest of this chapter focuses on illustrating the practical implementation of the IoT analytics lifecycle as part of the VITAL internet-of-things (IoT) platform for Smart Cities, which has been already introduced in Chapter 4. Furthermore, it presents practical examples associated with the deployment and operationalization of advanced IoT analytics, over footfall datasets collected from a smart city. IoT Data Collection.

In terms of IoT data collection, VITAL enables the collection of data from heterogeneous IoT systems, notably systems that have been developed and deployed independently in the scope of a smart city. To this end VITAL defines the PPI (Platform Provider Interface) abstract interface, which enables the unification of data from diverse systems in terms of their format. In particular, VITAL enables the collection of IoT data from different systems and data sources as soon as the latter implement the PPI interface.

The VITAL platform provides also the means for managing, registering and de-registering IoT systems in its platform, based on PPIs. Furthermore, the

PPIs enable the data collection and retrieval based on a JSON-LD format, which facilitates the semantic unification of data streams from different systems and data sources. This boosts the application of IoT analytics over diverse IoT systems, through alleviating the semantic heterogeneity of the various streams. Hence, the VITAL platform addresses the variety of IoT data streams both in terms of their formats and in terms of their semantics.

9.1.1 IoT Data Analysis

VITAL supports the storage and processing of the semantically unified data within a datastore, thus facilitating IoT data analytics over data stemming from multiple data sources and systems. The VITAL datastore is supported by a NoSQL database. The VITAL platform offers a wide range of data processing functions over this datastore, including:

- Dynamic data discovery based on criteria such as sensor type and location.
- Filtering on specific data attributes and on the basis of appropriate thresholds for each attribute.
- Complex Event Processing towards producing events based on information contained in multiple IoT streams.

Moreover, VITAL supports the data analysis phase through its integration with libraries of the R project. The latter libraries enable the execution of machine learning schemes, such as regression, classification and clustering.

9.1.2 IoT Data Deployment and Reuse

The VITAL platform enables the deployment of data processing algorithms over semantically unified streams, which are stored in the DMS (Data Management Service) of the system. It also enables the management of registrations to the various IoT data sources (including IoT platforms and systems), which provide the data to the DMS. In this way, VITAL supports the deployment of IoT data and its integration within IoT analytics applications in-line with the third phase of the already presented IoT analytics lifecycle. The integration of IoT data within applications is supported in a way that enables the repurposing and reuse of IoT data across multiple applications. This is made possible on the basis of the semantic annotation of the IoT data streams according to the VITAL JSON-LD contexts.

9.2 Knowledge Extraction and IoT Analytics Operationalization

Based on the functionalities outlined above, VITAL can be used for knowledge extraction, as well as for the deployment and operationalization of IoT analytics. Prior to deploying an IoT analytics application, the discovery and testing of IoT data mining algorithms that are likely to extract the desired knowledge in a credible way is required. In this respect, IoT data mining (which is part of the second phase of the IoT Analytics lifecycle) is very similar to conventional data mining applications i.e. applications leveraging transactional data instead of IoT streams. Hence, mainstream models for data mining and analytics such as the Cross Industry Standard Process for Data Mining (CRISP-DM) Model for Knowledge Discovery [1] can be applied. CRISP-DM entails the following activities and phases:

- **Business Understanding**: This activity is the starting point of the process and refers to the need of understanding the business problem at hand. A sound understanding of the nature of the problem is a key prerequisite to identifying proper machine learning models.
- **Data Understanding**: This activity follows business understanding and aims at understanding the data. By inspecting and understanding the data experienced data scientists can gain valuable insights on the applicability of certain data mining schemes. Data understanding leads to identification of data patterns in the datasets, which can serve as basis for identifying candidate machine learning schemes.
- **Data Preparation**: This is tedious, yet indispensable task in the process, given that the collected datasets need to be transformed in a format appropriate for identifying appropriate data mining and machine learning models. In conventional data mining applications, the data preparation step involves multiple ETL (Extract Transform Load) processes. In the case of IoT analytics, data engineers will have to deal with a multitude of data sources and formats depending on IoT data streams involved. The data preparation process is in several cases tedious, as a result of the need to deal with heterogeneous data sources, formats and semantics. As already outlined, semantic interoperability solutions (such as VITAL) facilitate the data preparation process.
- **Modelling**: This activity leverages data sets collected from the IoT system in order to identify a proper machine learning scheme for the problem

at hand. This task is facilitate by data mining tools (such as RapidMiner[1] and Weka[2]), which can be used to produce a machine learning model (e.g., a classifier or an association) given a training dataset. The modelling phase interacts very closely with the data preparation phase in order to ensure that the available training datasets are appropriate for fitting the target/identified models.

- **Evaluation**: As part of this phase, the produced model is evaluated in terms of its efficiency as the latter is reflected in the speed of training, the speed of model execution, its noise tolerance, as well as its expressiveness and explanatory ability. The evaluation is based on metrics such as classification accuracy, errors in numeric prediction, lift and conviction measures and more. A validation datasets (which is different from the training dataset) is used in the scope of the evaluation process. In case of acceptable performance and accuracy, the data scientists and practitioners can move the model to deployment. However, in case of poor performance, the whole cycle (i.e. from business problem understanding to model evaluation) has to be repeated in order to identify a model that gives satisfactory results for the problem at hand.
- **Deployment**: Successful models (i.e. schemes providing acceptable performance for the business problem at hand) are deployed and operationalized. The VITAL platform and more specifically its development environment offers integration with the R project, as a means of easily programming and deploying IoT analytics schemes. Hence, identified data mining models and schemes (e.g., Bayesian classifiers, K-means clustering algorithms, logical regression schemes) can be flexibly programmed and integrated within an application workflow and accordingly deployed based on the VITAL middleware platform.

In following paragraphs we provide a concrete example of the knowledge extraction process based on IoT data streams.

9.3 A Practical Example based on Footfall Data

As a case-study of performing some useful analysis on sensor data using advanced data mining techniques, we analyze the Camden footfall dataset. The Camden dataset comprises a multi-dimensional time-series in a

[1]https://rapidminer.com/

[2]http://www.cs.waikato.ac.nz/ml/weka/

time-frame of 1 hour, for two months, of counts of people passing in front of 5 different cameras in Camden, London. The first two weeks of this data-set is depicted in Figure 9.1 where its periodic nature is revealed; in the figure, each time-series label InPlusOutX refers to the total number of people detected during the particular time-frame by camera X (in location X); the interested reader can find more information in http://www.springboard.info/service/service-display/visitor-counting.

By visually inspecting the time-series it is clear that the 4-th location (corresponding to the camera no. 4) is usually the busiest. It is also clear that there are dependencies between all locations (that can be confirmed by computing the R values for any pair of time-series components.) Logistic regression (as implemented in the Weka suite of tools [2]) does not provide much more insight into the nature of the relations between the time-series components. In order to obtain some more insight into the relations between the given time-series components, we have performed Quantitative Association Rule Mining (QARM), introduced in [3], using QARMA, a highly parallel/distributed algorithm for mining all non-dominated "interesting" quantitative association rules in multi-dimensional dataset [4]. Quantitative association rules are association rules defined over quantitative attributes which they qualify over certain intervals. We define a rule to be "interesting" when its support and confidence exceed 8% and 85% correspondingly. The notion of non-dominance in quantitative association rules is formally defined in [4], but intuitively, a rule r dominates a second rule s if whenever s fires (i.e. all its antecedents are satisfied), r also fires, the consequent part of r covers the consequent of s, and r has equal or higher support and confidence than s.

Running QARMA on the Camden dataset produces a total of more than 25.000 non-dominated rules, which entirely cover the dataset: every data point

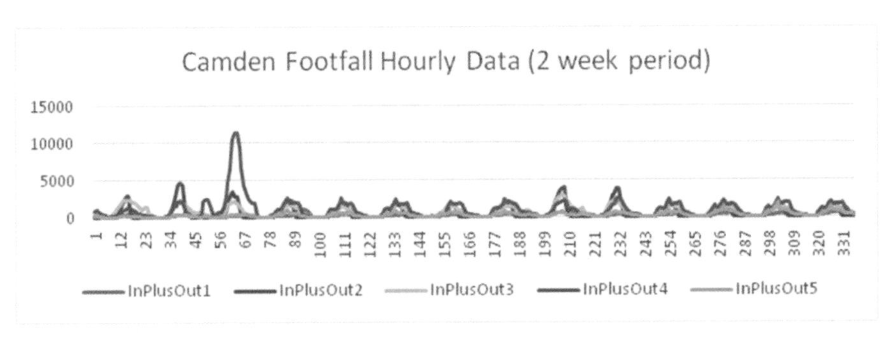

Figure 9.1 Camden footfall dataset.

in the dataset is covered by the application of at least one produced rule. The produced rules have a conviction [4] measure in the range [100.0%, 837.9%] which on average is 437.3%, while the lift of the rules [5] is in the range [1.0, 10.8], and averages at 2.87; both these measures indicate that the produced rules are far from being statistical flukes. Among the most interesting rules found, we list the following:

- Rule1: Time-of-day in [16:00, 19:00] $\longrightarrow$ InPlusOut1 $\geq$ 449 with support = 14.2%, confidence = 85.2%, conviction = 472.7%, lift = 2.82
- Rule2: InPlusOut3 $\geq$ 1874.0 $\wedge$ InPlusOut4 $\geq$ 423.0 $\wedge$ InPlusOut5 $\geq$ 262.0 $\longrightarrow$ InPlusOut2 $\geq$ 2051.0 with support = 8.6%, confidence = 85.1%, conviction = 579.9%, lift = 6.17
- Rule3: InPlusOut1 $\geq$ 265.0 $\wedge$ InPlusOut5 in [327.0, 1791.0] $\wedge$ Max{InPlusOut{1...5}} in [2309.0, 6554.0] $\longrightarrow$ InPlusOut2 $\geq$ 2334.0 with support = 8.9%, confidence = 85.06%, conviction = 603.7%, lift = 8.65
- Rule4: InPlusOut2 $\geq$ 2298.0 $\longrightarrow$ InPlusOut1 $\geq$ 365.0 with support = 8.811475409836065%, confidence = 85.4%, conviction = 415.85%, lift = 2.16

The first rule for example, shows that the camera in location 1 (InPlusOut1) whose average footfall is just under 336, in the 3-hour afternoon period between 16:00–19:00 increases above 449 regardless of the traffic in the other cameras or day of month.

The 2nd rule states that when locations 3, 4, and 5 are above certain footfall thresholds, then it is the second location that becomes the most crowded. This particular rule has among the highest conviction and lift rates, making it a very statistically significant and interesting rule.

The 3rd rule states that when footfall in location 1 is above a certain threshold, location 5 is within certain limits and the maximum of all locations is within certain limits as well, then location 2 exceeds its average value by more than 3 times (the average footfall measured by camera in location 2 is around 706).

Finally, the 4th rule provides an association between the footfall in location 2 and location 1, showing that when the footfall in location 2 is above a threshold that is (very significantly) above its average value, then the footfall in location 1 also increases above its average value. However, this last rule has a lift value of 2.16 and is thus not as strong as the previous three rules.

The produced quantitative rules fully describe the dataset, and show significant associations between the measured values of the time-series components; they also have the extra advantage of showing associations at "corner cases", that is they show what happens in one component when some other components significantly exceed their expected values, in fully quantifiable ways.

Acknowledgement

Part of this work has been carried out in the scope of the VITAL project (www.vital-iot.com), which is co-funded by the European Commission in the scope of the FP7 framework programme (contract No. 608662).

References

[1] Wirth, Rüdiger, and Jochen Hipp. *CRISP-DM: Towards a standard process model for data mining*. Proceedings of the 4th international conference on the practical applications of knowledge discovery and data mining. 2000.

[2] I. H. Witten, E. Frank, M. A. Hall, "Data Mining: Practical Machine Learning Tools & Techniques", Morgan Kaufmann, 3rd edition, Burlington, MA, 2011.

[3] G. Piatetsky-Shapiro, "Discovery, analysis, and presentation of strong rules", In: Piatetsky-Shapiro G., and Frawley, W. J. (eds), Knowledge Discovery in Databases, AAAI/MIT Press, Menlo Park, CA, 1991.

[4] I. T. Christou, E. Amolochitis, Z.-H. Tan, "QARMA: A Parallel Algorithm for Mining All Quantitative Association Rules and Some of its Applications", Under Review, 2016.

[5] M. Hahsle, B. Grün, K. Hornik: "arules – A computational environment for mining association rules and frequent item sets", Journal of Statistical Software 14(15):1–25, 2005.

10

Ethical IoT: A Sustainable Way Forward

Maarten Botterman

GNKS Consult NV, Netherlands

10.1 Introduction

The Internet of Things (IoT) is rapidly developing, primarily driven by businesses that see opportunities for profit through new business and business models. Other key players include public administrations and non-profit institutions that see IoT as an opportunity to address societal challenges in effective ways that were not reasonably available, before. Good public use of IoT can pay off rapidly, and lead to the ability of serving citizens and businesses better, at lower costs, in more inclusive ways. The first applications in the public domain show promising results – and it is still early days.

Evolution from machine to machine technology to the growing IoT networks raises challenges at every level that can become barriers to adoption when not addressed. New masses of data are generated by our Things and then shared between objects. Smart algorithms can combine this information with masses of very diverse sources such as social media, Open Data, traffic data, etc., leading to a world where BigData are more and more used to guide our decisions.

This world with "many eyes" (all the sensors in IoT and traffic data that register what happened where when and who/what was involved) and a wealth of (Big) data that can be combined and analyzed using smart algorithms is a world in which the old methods of privacy protections often fail [1]. Many notions of privacy rely on informed consent for the disclosure and use of an individual's private data. Data has become a resource that can be used and reused, often in ways that were inconceivable at the time the data was collected.

IoT has become a real game changer in this as sensors complement the data that were already generated, stored and shared before with new data,

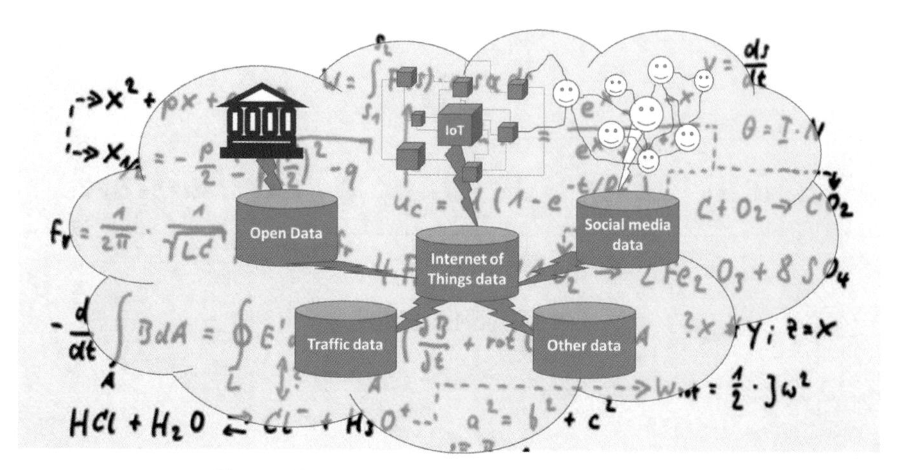

Figure 10.1 From IoT to BigData and analytics.

increasingly filling the gaps in digital representation of the (physical) world and what is happening on its surface. Not only that: IoT offers the opportunity to collect high value data – very focused on what the investing organisations want to know, and relatively well structured, as real-time as needed, and in context [2]. How do we adapt to this new reality where almost everything can eventually be captured in digital form (Section 10.2)? And what is needed most in order to create an environment that fosters positive evolution of the IoT, allowing us as businesses and society to benefit fully, without having to be afraid for the consequences (Section 10.3)? The conclusion (Section 10.4) is very much in line with the words of Commissioner Ansip: "Trust is a must".

10.2 From IoT to a Data Driven Economy and Society

It is clear: in terms of pervasiveness, IoT has already contributed to the emergence of a society in which almost everything is or can be monitored. It is not new, nor can further roll-out be stopped. What is new, is the enormous amount of "Things" that are now connected to the Internet and that are collecting, storing and sharing information . . . and the further rapid growth of deployment of more "Things"[3] and the increasing ability of actors to access and analyze data generated by IoT and many other sources.

Now: whereas the levels of monitoring are very high and well beyond the imagination of George Orwell [4] in terms of what technically is possible, in Europe trust in government and society has remained at a relatively high level.

When Snowden revealed, starting in June 2013, some evidence reflecting the pervasiveness of monitoring through numerous global surveillance programs [5], many of them run by the NSA (National Security Agency) and the Five Eyes[1] with the cooperation of telecommunication companies and European governments, this resulted in widely expressed concern and even outrage by the general public, civil society and politicians.

This led to a global discussion making clear that monitoring is a necessity, yet should be proportional, and not take place at all costs, and a balance is yet to be found. This results in a discussion that will continue to stretch over the decades to come.

Overall, it is noted also by the European Parliament that surveillance and collection of data should be proportional and justified, noting that new legislation is underway in multiple EU member states that would allow broad collection of data and tapping of internet communications: also including IoT [6].

Within this setting, the discussion in Europe about privacy and data protection is finding its way, moving from a Directive on Data protection and privacy towards European legislation that will come into full force in May 2018. The reform is to strengthen individual rights and tackle the challenges of globalisation and new technologies, and "simplify" compliance by being applicable law in all EU member states, whereas the Data protection Directive originating from 1995 was applied by national governments in similar but not always the same way.

When the original Data Protection Directive was developed and agreed in 1995, the Internet was by far not as important as today, and nobody had even mentioned the term "Internet of Things" yet. A review of the 1995 Directive in 2009, sponsored by the UK Data Privacy Authority, already noted that new developments like IoT, data mashups and data virtualization are new challenges that had to be met [7]. The reform that led to the new General Data Protection Regulation (further: GDPR) has been under way since 2011 and culminated in a Proposal to Council and Parliament by the European Commission on 25 January 2012. This proposal was approved by the European Parliament in March 2014, and has now been finalized and ratified by Parliament and Council to come into force in May 2018.

---

[1]"Five Eyes", often abbreviated as "FVEY", refer to an intelligence alliance comprising Australia, Canada, New Zealand, the United Kingdom, and the United States that was formed. These countries are bound by the multilateral Agreement, a treaty for joint cooperation in signals intelligence.

With this, it should be noted that the work has not been completed. When this law was set up in outline in 2011, "BigData" was not yet an issue widely recognized, in that year new in the Gartner Emerging Technologies Hype Cycle.

Today, we know that BigData, and BigData analytics, fundamentally challenge the concept of "personal data" as through BigData analytics data that in isolation do not relate to persons often can be related to persons when combined with other data.

The 2014 Opinion from WP29[2] on IoT recognises the value of IoT, as well as the potential intrusions it can generate to privacy. In this Opinion, statements are made that alarmed businesses around the world now asking for guidance to the European Data Protection Supervisor, as what is suggested may put a lock on many current developments in the field.

In 2015, a Court Ruling by the European Court of Justice in the case of Maximillian Schrems versus the Irish data protection commissioner regarding the right of Facebook to transfer data to servers located in the USA under the Safe Harbour scheme further led to uncertainty about the legal situation. On 6 October 2015, the Court declared the Safe Harbour Decision invalid, as the protections under the Safe Harbour scheme provided by the US Authorities had proven to be inadequate, in particular because "*the scheme is applicable solely to the United States undertakings which adhere to it, and United States public authorities are not themselves subject to it. Furthermore, national security, public interest and law enforcement requirements of the United States prevail over the safe harbour scheme, so that United States undertakings are bound to disregard, without limitation, the protective rules laid down by that scheme where they conflict with such requirements*" and because for non-US citizens there is no opportunity to redress: "*legislation not providing for any possibility for an individual to pursue legal remedies in order to have access to personal data relating to him, or to obtain the rectification or erasure of such data, compromises the essence of the fundamental right to effective judicial protection*".[3]

[2]The Article 29 Data Protection Working Party was set up under the Directive 95/46/EC of the European Parliament and of the Council of 24 October 1995 on the protection of individuals with regard to the processing of personal data and on the free movement of such data. It has advisory status and acts independently.

[3]ECJ ruling in case C-362/14 Maximillian Schrems vs Data Protection Commissioner, 6 October 2015, http://curia.europa.eu/jcms/upload/docs/application/pdf/2015-10/cp150117en.pdf

Currently, IoT providers such as the globally popular "Nest" smart meters and smoke detectors, owned by Google, still refer to "Safe Harbour Agreement" protection of personal data.[4] Whereas Nest explicitly commits to a number of privacy measures, it should be noted that today (2015) redress is thus not possible when a European citizen considers her or his privacy right to be violated and their data are kept in database physically outside of Europe. This is also true for companies such as Younqi (health bands) and many others that collect data and store them on US servers. The measures currently proposed by the European Commission to replace "Safe Harbour", known as "Privacy Shield", have not been accepted, yet, and await an advice from WP29. Note that the EU-U.S. "Privacy Shield" imposes stronger obligations on U.S. companies to protect Europeans' personal data and requires the U.S. to monitor and enforce more robustly, and cooperate more with European Data Protection Authorities. It also includes written commitments and assurance regarding access to data by public authorities. It should also be noted that this new agreement has not been tested in Court, yet, and until this is done and ruled to be a "valid agreement" uncertainty about this new protection remains.

Businesses are looking for guidance, as BigData is a subject of interest to many, and companies around the world are looking into the opportunities offered by BigData, data generation, collection, and analytics. IoT is a major driver in this, as "connected Things" will generated endless streams of data that will be captured and used. According to the European Data Protection Supervisor Peter Hustinx [8]: "*If BigData operators want to be successful, they should . . . invest in good privacy and data protection, preferably at the design stages of their projects*".

With this, he recognises the important of "soft law" at this point [9]. Investing in good privacy and data protection should be core in the innovation, development and deployment of IoT, and probably a pre-condition for European (co-)sponsored research. A way forward could include the habit/obligation of a Privacy Impact Assessment in every stage of design of new IoT products and services.

In his published Opinion on Digital Ethics [10], the European Data Protection Supervisor (EDPS) refers to Article 1 of the EU Charter of Fundamental Rights: '*Human dignity is inviolable. It must be respected and protected.*' From that position he further explains that: "*In today's digital environment, adherence to the law is not enough; we have to consider the ethical dimension of data processing.*"

[4]https://nest.com/legal/privacy-statement-for-nest-products-and-services/

It is in line with this that some projects funded by the European Commission are looking very carefully at the issue of privacy protection and the idea of limiting the amount of information available to each entity. In general, the key issue to take into account while discussing privacy has to do with the integration of information from different sources. While a single stream of data might not contain enough information to invade the privacy of the user, it is recognized that the correlation and concurrence of information at an entity can lead to privacy considerations that were unthinkable only looking at the individual sources.

While the user is ultimately responsible for the data it allows to escape in the open, a modern individual that works and lives with current technologies cannot keep up with the types and amount of information being "leaked" by applications and websites. It is, therefore, for an individual virtually impossible to design privacy policies that are permissive enough to allow for services to work, while at the same time, restrictive enough that the privacy of the user is not compromised. Any specific harm or adverse consequence is the result of data, or their analytical product, passing through the control of three distinguishable classes of actor in the value chain [11]:

1. Data collectors, who may collect data from clearly private realms, from ambiguous situations, or data from the "public square," where privacy – sensitive data may be latent and initially unrecognizable;
2. Data analyzers, who may aggregate data from many sources, and they may share data with other analyzers creating uses by bringing together algorithms and data sets in a large – scale computational environment, which may lead to individuals being profiled by data fusion or statistical inference;
3. Users of analyzed data generally have a commercial relationship with analyzers; creating desirable economic and social outcomes, potentially producing actual adverse consequences or harms, when such occur.

As the complexity increases through technology, we will depend on technology to deal with it. It is crucial that automated and self-configuring solutions are offered that analyze the type and amount of information given away for a specific user and configure the appropriate number of policies to ensure that the level of security and privacy desired by the user is kept untouched. This goes beyond mere regulatory actions and requires robust and flexible technology solutions that work under very different conditions, and that are backed by legislation to ensure that abuse of technologies or data is subject to redress and legal action.

10.3 Way Forward with IoT

The Internet of Things and its underlying data streams shape an important part of that transformation by collecting and sharing data from a rapidly increasing amount of objects that are digitally connected in our lives, ranging from our cars to smart TVs, smart homes to smart cities, as well as natural disaster warning systems and air quality sensors network.

Drivers for IoT introduction include the need to address societal challenges in efficient ways, and to grab business opportunities that often come with new business models. There is no way back: the “promises” of IoT make further development unstoppable. Data generated and shared by objects connected through the Internet, combinable using smart algorithms, lead to a world in which privacy is getting a new meaning and where good security is more important than ever. IoT, in combination with BigData and data analytics in particular as an enabler of high quality real-time data provider, has become a real game changer.

Governments at all levels are confronted with this, and need to find responses, soon. Societal challenges need to be dealt with effectively, using less money and relying more on active participation of citizens and businesses, yet this cannot go at cost of a society we want in terms of trust.

IoT is currently mainly driven by business opportunity considerations and technology push, yet it is clear that people are waking up and become concerned on where this takes us. Consumers and citizens have to become involved in developing a “future we want” in which there is “respect for human dignity” as well as individual choice.

During meetings within the European IoT and Future Internet research community and the recently launched public private partnership *Alliance for Internet Of Things Innovation* (AIOTI), and in global forums such as the Internet Governance Forum’s *Dynamic Coalition of the Internet of Things* (IGF DC IoT) and EuroDIG, these issues have been at the center of a dialogue between public, private, and civil sector stakeholders. There is an ongoing need to protect the public interest as well as to create space for innovation and experimentation using IoT products and services within the current and developing legal frameworks. To find this balance requires the active, well informed involvement of public authorities.

IoT can be used for many different things in many different ways, and practical experimentation in an ecosystem in which all stakeholders are involved will help understand the impacts of the IoT more profoundly than its technological specifications alone. In order to be “trusted” by its users, IoT will need to offer:

- Meaningful transparency – what is happening;
- Clear accountability – who takes responsibility;
- Real choice – "all or nothing" is not good enough.

Dialogues at global level have led to the insight that IoT needs to "go ethical". What this means exactly, and how it can work is still to be determined, with all stakeholders around the table. One thing is clear: we cannot continue to count on "adherence to the law". A good first step has been made with the draft declaration by the *2015 IGF Dynamic Coalition on Internet of Things Good Practices Policies* which can be found on the website of the IGF.[5]

10.4 Conclusions

It is clear that IoT and BigData have changed our ability to protect data to be related to individuals. At the same time, this doesn't mean we should give up on the right to privacy. To quote the EDPS: "*there are deeper questions as to the impact of trends in data driven society on dignity, individual freedom and the functioning of democracy*." And to quote the US President's Council of Advisors on Science and Technology [11]: "policy focus primarily on whether specific uses of information about people affect privacy adversely. It also recommends that policy focus on outcomes, on the "what" rather than the "how," to avoid becoming obsolete as technology advances. The policy framework should accelerate the development and commercialization of technologies that can help to contain adverse impacts on privacy, including research into new technological options. By using technology more effectively, the Nation [USA] can lead internationally in making the most of BigData's benefits while limiting the concerns it poses for privacy."

As Commissioner Ansip stated in his speech (spoken word) during the Net Futures conference in Brussels on 20 April 2016: "Trust is a must" for whatever we do on our way forward.

No stakeholder can do this alone. Businesses need to invest, governments need to protect the public interest which includes protection, ensuring redress and choice, the technical community needs to design and develop new and better approaches, and users need to be aware and "steer" investments and developments through conscious use.

Time to make technology work for us in a way that people can trust these technologies is now. Let's make sure we reflect our awareness of and

[5]http://review.intgovforum.org/igf-2015/dynamic-coalitions/dynamic-coalition-on-the-internet-of-things-dc-iot-4/

commitment to this ethical side in every step we do when developing and deploying new technologies and services that collect, store, share, protect and act on data.

For those of us who have read Asimov's book "*I, Robot*" [12]: remember the Three Laws of Robotics which in fact could relate to all intelligence developed, and apply them to whatever connected intelligence you work on – no harm to be done to people. Indeed, Isaac Asimov, describes the three laws that set the way forward for robots: 1st Law: A robot may not injure a human being or, through inaction, allow a human being to come to harm; 2nd Law: A robot must obey the orders given it by human beings except where such orders would conflict with the First Law; 3rd Law: A robot must protect its own existence as long as such protection does not conflict with the First or Second Laws.

References

[1] Helen Rebecca Schindler, Jonathan Cave, Neil Robinson, Veronika Horvath, Petal Jean Hackett, Salil Gunashekar, Maarten Botterman, Simon Forge, Hans Graux. *Europe's policy options for a dynamic and trustworthy development of the Internet of Things*. Report for the European Commission under SMART 2012/0053. June 2013.

[2] Tableau Software. *Top 8 Big Data trends for 2016*. Report.

[3] Maarten Botterman. *Opening towards a new reality*. Policy paper on IoT Future Technologies for the European Commission DG CONNECT. Rotterdam, April 2015.

[4] George Orwell. 1984. *Secker and Warburg*. June 1949.

[5] Glenn Greenwald, Ewen MacAskill and Laura Poitras. *Edward Snowden: the whistleblower behind the NSA surveillance revelations*. the Guardian, June 2013, http://www.theguardian.com/world/2013/jun/09/edward-snowden-nsa-whistleblower-surveillance.

[6] Niels Muiznieks. *Europe is Spying on you*. NY Times http://www.nytimes.com/2015/10/28/opinion/europe-is-spying-on-you-mass-surveillance.html?_r=0, October 2015.

[7] Neil Robinson, Hans Graux, Maarten Botterman & Lorenzo Valeri. *Review of the EU Data Protection Directive*. prepared for the Information Commissioner's Office, TR-710-ICO. Cambridge, May 2009.

[8] Mark Say. *Big data needs big guidance*. Financial Times. Retrieved http://www.ft.com/cms/s/0/fab4bae8-7f88-11e4-86ee-00144feabdc0.html#axzz3O8I1GAvc on 2015.01.07, December 2014.

[9] RAND. *Europe's policy options for a dynamic and trustworthy development of the Internet of Things*. June 2013.
[10] European Data Protection Supervisor. Opinion 4/2015. *Towards a new digital ethics: Data, dignity and technology*. EDPS, September 2015.
[11] Executive Office of the [USA]. *Big Data: Seizing Opportunities, Preserving Values*. President PACT report. May 2014.
[12] Isaac Asimov. I, Robot. Gnome Press December 1950.

Epilogue

IoT Analytics is without a shadow of a doubt one of the most important elements of the IoT computing paradigm, which will probably contribute the largest portion of IoT's business value in the years to come. In this book we have introduced the main challenges that are associated with IoT analytics systems and application, along with the main technological elements that comprise non-trivial IoT applications. IoT analytics systems are essentially BigData systems, which usually have to collect, process and analyze heterogeneous high-velocity streams. Therefore data streaming infrastructures and semantic interoperability technologies for IoT streams are among the main pillars of IoT analytics systems and applications. Furthermore, cloud computing infrastructures are an integral component of IoT analytics, since they provide the capacity, scalability and elasticity required in order to deal with large amounts of IoT data.

In the scope of the book we have presented popular middleware infrastructures for handling distributed streams with high ingestion rates, along with tools and techniques for semantic modeling and interoperability of highly heterogeneous data streams stemming from different sources and devices. We have also illustrated the integration of IoT data streams in the cloud and the role of cloud computing technologies in IoT analytics. The presented infrastructures and technologies provide the reader with a sound understanding of what engineers, researchers and practitioners can nowadays use in order to implement, deploy and operate IoT analytics applications. Most of the presented systems and technologies for IoT analytics are open source, thus providing a good starting point not only for practitioners wishing to deploy the systems, but also for students and researchers wishing to explore and learn IoT analytics.

Along with IoT analytics technologies, the book has also presented a set of indicative practical deployments of IoT analytics systems in areas such as smart buildings, smart cities and crowd analytics. As part of the presentation of these applications, the use of the earlier presented technologies has been substantiated in the scope of practical systems. Moreover, the presentation of these practical systems has illustrated the importance of machine learning and data mining technologies in the data analysis process. As part of the practical case studies we have therefore illustrated how different machine learning techniques can be developed, tested, evaluated and ultimately deployed in the context of an IoT system.

The book is certainly a good starting point understanding the scope of IoT analytics and the tools that are already available in order to build IoT applications. Neverhteless, the presented systems and applications are only the tip of the iceberg. In the coming years, systems with significantly increased sophistication and complexity will emerge, far beyond the collection, homogenization and mining of IoT streams. The emergence of such systems will drive a radical shift of IoT's focus from the "best IoT product" (e.g., the best smart-phone or wearable IoT device) to the "best IoT service" (e.g., personalized context-aware recommendations for fitness, training and a healthy lifestyle). This shift will be propelled by IoT analytics, as it will be the collection and processing of IoT data that will enable the creation of human centic IoT services in consumer markets, as well as the creation of after sales programs in the market of industrial goods and services. This shift will signal an unprecedented revolution that could completely change our everyday living. Moreover, it will also come with a shift in IoT analytics tools and techniques. This anticipated revolution however is not bound to change the value of the present book as the IoT analytics building blocks that have been presented in earlier chapers will form the foundation for the development of the novel revolutionary solutions. We really expect this book to help its readers not only to familiarize themselves with mainstream IoT analytics technologies but also to remain equipped for the rising revolution of IoT and IoT analytics.

Index

A
Analytics 1, 5, 11, 15
Apache Storm 9

B
BigData 3, 11, 17, 81

C
Change point Analysis 52
Cloud Computing 3, 27, 99, 217
Cloud Services 84, 100, 105
CRISP-DM 233

D
Data Analysis 8, 14, 19, 41, 87
Data Analytics 15, 81, 101, 139
Data Mining 8, 147, 233
Data Protection 241

E
Edge Analytics 216, 227
Energy Efficiency 169, 184, 192
Ethics 243
Event retrieval 48, 51, 57

G
Gateway 14, 21, 83, 143

I
IDE 83
Information Retrieval 47, 57
Integrated Development Environment 82, 84
Internet of Things 3, 11, 39, 142
Internet-of-Things analytics 207
Internet-of-Things platform 12, 25, 43, 156
Interoperability 5, 14, 34, 84
IoT 3, 5, 11, 15
IoT Analytics 3, 5, 7, 11
IoT application development 6, 17, 81, 100
IoT Platform 12, 15, 43, 156

L
Learning-to-Rank 47, 69, 72
Linked Data 14, 43, 141, 150
Location Based Social Network 63
LoRa 12, 21
Low-Power 20

N
Node-RED 25, 82, 87, 90

O
Open Source 9, 14, 19, 22
OpenIoT 101, 108, 113, 123
Optimization 3, 11, 35, 102

P
PaaS 12, 24, 100, 216
Privacy 6, 29, 185, 225

R
Real-time search 41, 73
Reasoning 43, 139, 145, 151
Regulation 241
Resource Management 106, 118
Rules Mining 8

S
Scheduling 107, 112, 216, 223
Security 6, 29, 50, 173
Semantic Interoperability 5, 84, 96, 151
Semantic Reasoning 145, 218

Semantic Sensor Networks (SSN) 101, 142
Sensor Streams 42, 46, 63
SMART 41, 45, 63, 73
Smart Buildings 167, 169, 178, 181
Smart Cities 3, 20, 84, 167
Smart City 86, 141, 208, 213
Social Media 39, 44, 49, 239
Social Sensing 39, 48, 64, 73
SPARQL 101, 110, 117, 130
Streams 3, 6, 29, 41

T

Terrier 44

U

Utility-Driven Models 138

V

Variety 5, 10, 16, 19
Velocity 5, 16, 81, 96
Venue recommendation 40, 63, 68, 70
Veracity 5, 8, 16
VITAL 81, 84, 87, 96
Volume 5, 13, 19, 44

W

WAN 20, 21

About the Editor

John Soldatos holds a Ph.D. in Electrical & Computer Engineering from the National Technical University of Athens (2000) and is currently Associate Professor at the Athens Information Technology (2006–present) and Honorary Research Fellow at the University of Glasgow, UK (2014–present). He was also Adjunct Professor at the Information Networking Institute (INI) of the Carnegie Mellon University, Pittsburgh, USA (2007–2010). In the period 2012–2014 he served as the coordinator of the working group "*IoT identification, naming and discovery*" of the European Internet-of-Things Research Cluster (IERC). Dr. Soldatos has played a leading role as software engineer, systems architects, technical project manager or consultant, in the successful delivery of more than fifty (50) (commercial, industrial, research and consulting) projects, for both private & public sector organizations, including complex integrated projects on Internet-of-Things and Enterprise Data Analytics. He is co-founder of the open source platforms OpenIoT and AspireRFID, while he has been the technical manager of several EC co-funded projects on the Internet-of-Things, including the FP7 ASPIRE, OpenIoT and SMART projects. He has published more than 180 articles in international journals, books and conference proceedings, while he has written numerous posts and articles for blogs, newspapers and magazines. He has also significant academic teaching experience, along with experience in executive education and corporate training. During July 2014–March 2016 he has served as a member of the European Crowdfunding Stakeholders Forum.

About the Authors

Amelie Gyrard holds a Ph.D. from Eurecom since April 2015 with highest honors and she has been selected as one of the 10 nominees for Best Ph.D. thesis Price – Telecom ParisTech 2015 – France. She designed and implemented the Machine-to-Machine Measurement (M3) framework (http://sensormeasurement.appspot.com/) under the supervision of Prof. Christian Bonnet (Eurecom) and Dr. Karima Boudaoud (Universite Nice Sophia Antipolis, I3S, CNRS). The title of her dissertation is "Designing Cross-Domain Semantic Web of Things Applications". An entire workflow to semantically annotate and unify heterogeneous IoT data, a necessary step to reason on data to deduce meaningful knowledge and query enriched data. This framework hides the complexity of using semantic web technologies to the developers by generating IoT application templates. Moreover, it shows the importance to combine heterogeneous applicative domains. She also disseminated her work in standardizations such as ETSI M2M, oneM2M (working group on management, abstraction and semantics) and W3C Web of Things. She is also reviewer of international conferences and journals such as Semantic Web Journal, Sensors Journal, IEEE Internet of Things Journal. She is also highly involved within the "Semantic Web of Things" tutorial at ISWC 2016.

Dr. Antonio F. Skarmeta received the M.S. degree in Computer Science from the University of Granada, Spain, and the B.S. (Hons.) and the Ph.D. degrees in Computer Science from the University of Murcia. Currently, he is a Full Professor in Dept. of Information and Communications Engineering at the same university. He is involved in numerous projects, both European and National. Research interests include mobile communications, artificial intelligence and home automation.

Aurora González-Vidal graduated in Mathematics at the University of Murcia in 2014. In 2015 she got a fellowship to work in the Statistical Division of the Research Support Services, where she specialized in Statistics and Data

Analyisis. During 2015, she started her Ph.D. studies in Computer Science, focusing her research in Data Analytics for Energy Efficiency.

Bin Cheng is a senior researcher in the group of Cloud Services and Smart Things at NEC Laboratories Europe. He has been working on big data analytics platforms for Smart Cities and machine learning based task management for cloud services. He led the design and implementation of a big data and analytics platform for the Santander city, which has been in production use. His expertise includes scalable NoSQL databases like CouchDB, big data processing platform like Spark and Storm. His research has focused on big data analytics and edge computing platform for IoT systems. He had publications at many system conferences, such as IEEE CLOUD, EuroSys, NSDI, and IMC. His Ph.D. dissertation won the outstanding doctoral dissertation award of China Computer Federation in 2010.

Dr. Congduc Pham is a Professor of Computer Science at the University of Pau (France) in the LIUPPA laboratory in which is served as director from 2006 to 2009. He obtained his Ph.D. in Computer Science in 1997 at the LIP6 Laboratory, University Pierre and Marie Curie. He also received his Habilitation in 2003 from University Claude Bernard Lyon 1. From 1998 to 2005, he was associate professor at the university of Lyon, member of the INRIA RESO project in the LIP laboratory at the ENS Lyon. His current research interests include sensor networking, Internet of Thing, congestion control and QoS for grid/cloud computing. He served as Guest Editor for Future Generation Computer System, Annals of Telecom and IJDSN. He serves as Program Vice-Chair for IEEE GreenCom 2010, workshop Co-Chair for IEEE WiMob 2011 and Track Chair for IEEE WiMob 2015. He has published more than 120 papers in international conferences and journals, has been reviewers for a numerous number of international conferences and magazines, and has participate to many conference program committees. He is a member of IEEE.

Craig Macdonald is a Lord Kelvin Adam Smith Research Fellow in Sensor Systems at the University of Glasgow. He has a background in information retrieval, and has over 100 publications in areas such as venue recommendation, efficient & effective search engines, learning-to-rank, enterprise search, diversification and online evaluation. He has a track record of collaboration with industry and the public sector. He is lead maintainer of the Terrier search engine platform, and was co-investigator for the University of Glasgow within the SMART FP7 project.

Dyaa Albakour is a Data Scientist at Signal Media in London, United Kingdom. He was previously a post-doctoral researcher at the University of Glasgow, where he worked upon smart cities projects, including the SMART FP7 project and the integrated Multi-media City Data project of the Urban Big Data Centre. His major research interest is using data-driven approaches for providing users with information access methods in news media, enterprises and smart cities. He holds a Ph.D. from the University of Essex since 2012.

Dr. Fang-Jing Wu is a Research Scientist at the group of Cloud Services and Smart Things at NEC Laboratories Europe. She is working on crowd mobility analytics at NEC Laboratories. She was awarded Ph.D. degree in Computer Science from the National Chiao Tung University, Taiwan, in 2011. She was a visiting researcher at Imperial College London from 2010–2011 and was a Research Fellow in Nanyang Technological University in 2012. She was a scientist at Institute of Infocomm Research (I2R), A*STAR, Singapore from 2013 to 2015. Fang-Jing was awarded Google's Anita Borg Memorial Scholarship in 2011 (the winner in Taiwan in 2011) and has published papers in several journals, conferences, and demo papers including ACM trans. on Sensor Networks, IEEE Sensors Journal, Ad Hoc Networks, IEEE Communications Letters, Wireless Communications and Mobile Computing, Pervasive and Mobile Computing, ICC, SenSys, ICPP, MASS, WCNC, Globecom, and SIGCOMM. Fang-Jing also contributes to academic communities in many ways as serving as TPCs of conference, organizing committee of conferences and workshops, and editorial board of International Journal of Ad Hoc and Ubiquitous Computing. Her current research interests are primarily in pervasive computing, wireless sensor networks, cyber-physical systems, participatory sensing, and mobile sensing.

Dr. Fernando Terroso-Sáenz graduated from University of Murcia with a degree in Computer Science in 2006. He also received the master degree in Computer Science at the same university in 2010. Then, he finished his Ph.D. at the Dept. of Communications and Information Engineering in 2013. Since 2009, he have been working as a researcher in this group. His research interests include Complex Event Processing (CEP), Ubiquitous Computing and Intelligent Transportation Systems (ITSs).

Flavio Cirillo is a Research Scientist of the Cloud Services and Smart Things group at NEC Laboratories Europe, Germany. His research focus is in the Internet-of-Things Platforms and Analytics especially in the topic of Smart City. He is currently the one of the main developer and maintainer of the IoT Backend layer of the IoT Architecture of FIWARE. He has worked in IoT related European research project such as FIWARE and FIESTA-IoT. He has obtained a Master degree in Computer Engineer at the University of Naples Federico II in 2014.

Iadh Ounis is a Professor of Information Retrieval at the University of Glasgow, with over 150 publications in information retrieval. He has a diverse set of funding from RCUK, EU, industry and public sector, including being the Principal Investigator for the University of Glasgow within the SMART FP7 project. Iadh is Director of Knowledge Exchange at the Scottish Alliance of Informatics and Computing Science Alliance (SICSA), and board member of the SFC Data Lab Innovation Centre. He specialises in large-scale search systems, and is a founder of the Terrier.org search engine platform.

Dr.-Ing Abdur Rahim is a senior research staff in smart IoT group at CREATE-NET, Italy. He is project co-ordinator of WAZIUP (www.waziup.eu) an Open Sources IoT and Big data Platform for Africa. He is also the technical manager of iKaaS (www.iKaaS.eu) an IoT, big data and Cloud project funded by EU and Japan. He was also the project manager of EU large-scale integrated Internet of Thing (IoT) project iCore (www.iot-icore.eu) for empowering Internet of Thing through cognitive technologies. He serves as technical working group leader of several EU projects and European cluster activities. He is very active in IoT since 2006 starting with sensors network. He delivery talks in many conference, events, meeting, etc. Dr.-Ing Biswas organized many technical/panel session and workshop at different international events like ICC, CROWNCOM, ICT Mobile Summit, EU RAS, etc., He involved in many activities related to cognitive technologies, steering committee co-chair (CROWNCOM), management board member of COST-TERRA, Co-Chair of SIB European Alliance for Innovation.

Dr.-Ing Corentin Dupont is a researcher at CREATE-NET research center, Italy. He received a Master's degree in Electronics and IT Technology in 2003 from the Institut Superieur de l'Electronique et du Numerique, France and a Ph.D. from the University of Trento, Italy. He is involved in the project WAZIUP (www.waziup.eu), an Open Sources IoT and Big data Platform for Africa. He was involved in several other EU funded projects, such as FIT4Green, DC4Cities and iCore projects. His research interests are in the domain of Functional and Constraint Programming, Green Service Level Agreements, and in general in Cloud computing and IoT.

Prof. Ioannis T. Christou (male) holds a Dipl. Ing. degree from the National Technical University of Athens, Greece in Electrical Engineering (1991), and an M.Sc. and Ph.D. degree in Computer Sciences from the University of Wisconsin-Madison (U.S.A.) (1993, 1996). He also holds an MBA from the Athens MBA joint program of NTUA and the Athens University of Economics and Business. His research interests are in the areas of parallel and distributed computing and optimization, data mining and machine learning. Dr. Christou has more than 70 research publications in peer-reviewed journals and conferences. He has been a Sr. Systems Engineer and Sr. Developer at Transquest Inc. and Delta Technology Inc., an area leader in Knowledge & Data Engineering at Intracom S.A. and a Member of Technical Staff at Lucent Bell Laboratories Inc. Dr. Christou has been an adjunct Assistant Professor at the Computer Engineering and Informatics Dept. of the University of Patras (2003–2004), and an adjunct professor at the Information Networking Institute of Carnegie-Mellon University in Pittsburgh PA, USA, (2007–2011). Since 2004, he has been with Athens Information Technology where he is now a full Professor. In 2013 he co-founded IntelPrize, a BigData start-up company focusing in providing personalized pricing and recommendations using Big Data and Data Mining technologies, for which he obtained initial funding and served as company CTO until 2016.

Jarana Manotumruksa is a research student at the University of Glasgow. He attained his Bachelor (Honours) degree in Computing Science from the University of Glasgow in 2014. His research areas are recommendation system and deep learning algorithms that can effectively suggest venues to visit based on user's preferences as well as contextual factors affecting the user (e.g., time of the day, user's location).

Katerina Roukounaki is a researcher and software engineer. She has a degree in Computer Science, a M.Sc. in Information and Telecommunications Technologies, and she is now pursuing her Ph.D. in Programming Languages. Over the years, she has worked in several different fields (e.g., telecommunications, space, internet-of-things), used several different languages (e.g., Java, Erlang, Python, JavaScript), and developed different types of applications (e.g., web, desktop, mobile). She has been involved as a senior researcher in various EC co-funded research projects, while she is co-author of several publications.

Dr. M. Victoria Moreno received the B.S. (Hons.) and the M.S. degrees in Telecommunications Engineering in 2006 and 2009, respectively, both of them from the School of Telecommunication Engineering of Cartagena, Spain; and the Ph.D. degree in Computer Science in 2014 from the University of Murcia, Spain. Currently, she is a post-doctoral researcher of the Spanish Seneca Foundation. Research interests include data analysis and modelling, and energy efficiency in smart environments.

Maarten Botterman is an experienced strategic advisor with over 25 years of experience in dealing with strategic and practical issues relating information technology and the Internet. Having served in Dutch national government, the European Commission and for the European office of RAND Corporation he has dedicated his life to work in the public interest. He has a reputation as independent, trustworthy, out-of- the-box thinker who is able to look at issues from several perspectives with strong awareness of the bigger picture. In this he has an excellent grasp of the impact of technologies and the move towards data on business, including the aspects of cybersecurity and privacy and data protection related risks. Next to his work as strategic advisor he also serves as Director on the ICANN Board, and as Chairman of the Supervisory Board of NLnet Foundation. Since 2014 he is also Chairman of the IGF's Dynamic Coalition on the Internet of Things.

Dr. Martin Bauer is a Senior Researcher in the group of Cloud Services and Smart Things at the NEC Laboratories Europe. He is/has been working on the European research projects MobiLife, SPICE, MAGNET Beyond, SENSEI, IoT-A, IoT-i, FIWARE, SMARTIE and FIESTA on topics related to context aware systems and IoT architecture and platforms. He contributed to the definition of the NGSI context interfaces in OMA and he is currently active in oneM2M standardization, driving the work on semantics there. He has (co-)authored more than 40 technical papers and has also been active as peer reviewer and program committee member for several journals, conferences and workshops.

Martin Serrano is currently an Adjunct Lecturer and Research Fellow with the Insight Centre for Data Analytics (formerly DERI), National University of Ireland, Galway, Ireland. He has more than ten years of experience with information and communication technology application and research, at the technical and administrative management levels, within a wide range of European international and Irish national collaborative projects. He is leading research on cloud services and communication infrastructures management. He is the author of the book Applied Ontology Engineering in Cloud Services, Networks and Management Systems (Springer-Verlag, 2012). He is currently exploring semantic techniques applied to cloud systems by using autonomic management principles as an approach for producing cognitive applications capable of understanding service and application events to control cloud service lifecycle. He is also investigating open source large-scale self-organizing cloud environments for IoT applications in the Future Internet.

Nikolaos Kefalakis holds a Diploma in Electronic Computing Systems from TEI of Piraeus, a M.Sc. in Information Technology and Telecommunications from AIT (Athens Information Technology) and is currently a Ph.D. candidate at Aalborg University of Denmark. Since February 2008, he has been working as a research scientist in the AGC Group of AIT where he has been involved in FP7 research projects. Specifically in the context of ASPIRE and OpenIoT FP7 projects he is the manager, system architect and developer lead of the AspireRFID OS project (http://wiki.aspire.objectweb.org/) and OpenIoT OS project (https://github.com/OpenIotOrg/openiot). Mr. Kefalakis main area of technical expertise is IoT systems, Auto-ID Technologies (RFID, Barcodes...), Semantic Sensor Networks, Multitier architecture systems, RCP applications, Enterprise Systems, Embedded and Electronics digital systems.

Richard McCreadie is a post-doctoral researcher at the University of Glasgow. He attained his Ph.D. degree in information retrieval in 2012. He has a strong track record in big data and social media analytics. He is currently leading the research community effort on real-time search and summarisation of social media with colleagues from Microsoft and the University of Waterloo.
Salvatore Longo obtained a Master degree in Software Engineer from Federico II University of Naples, Italy, in February 2012. In November 2011 he joined NEC Laboratories Europe, Germany, and worked as a software engineer in the CSST group. In 2014 has been promoted as Research Scientist within NEC and his research focus was on Data Analytics and Human Behaviour Analysis using real time system for IoT. He has been involved in several EU projects related to the Internet of Things area like IoT-A, FI-WARE and MOBiNET. In May 2016 Salvatore left NEC Research Lab to join Bosch Automotive Service Solutions to work on telediagnostic systems.